Jigs & Fixtures for Limited Production

jigs & fixtures
for limited production
Harold Sedlik

Published by
Society of Manufacturing Engineers
Dearborn, Michigan
1970

Manufacturing Data Series

**Jigs and Fixtures for
Limited Production**

Library of Congress Catalog Card Number: 74-118842
International Standard Book Number: 0-87263-025-0

MANUFACTURED IN THE UNITED STATES OF AMERICA

To my wife

Candide

in appreciation of
her encouragement and support

PREFACE

The primary function of a jig or fixture is to aid in production and at the same time to reduce production costs. In the area of limited production, however, the use of jigs and fixtures presents a challenge, for the costs of these tools themselves may often exceed income from the product. Consequently, as the volume of production is reduced, tooling costs must also be reduced in proportion.

The purpose of this book is to guide the tool engineer in developing the most economical jigs and fixtures he can devise to confront the need for reduced costs in limited-production manufacturing. Throughout the text, emphasis is placed on the economics, design, and fabrication of inexpensive yet efficient jigs and fixtures — tools that will fulfill tooling requirements at a lower cost.

I regret that the fundamentals of jig and fixture design and of toolmaking could not be included in this book, but other authors in other books have covered these subjects in depth. I hope this book will suggest numerous ways of reducing tooling expenses and encourage the cost-conscious engineer to look beyond fundamental tooling in his search for those cost reductions.

HAROLD SEDLIK

West Hartford, Connecticut
September 30, 1970

contents

CHAPTER 1   INTRODUCTION 3
Definition of Terms 3
Jigs and Fixtures in Manufacturing... 3
Mass Production and Limited Production 4
Aspects of Limited Production..................................... 5
The Challenge to Tool Engineers 7

CHAPTER 2   TOOLING FOR LIMITED PRODUCTION 9
Manufacturing Costs 9
Tool Costs vs. Labor Costs 9
Presenting the Tooling Estimate 11
Manufacturing Without Special Tooling......................... 13
Evaluation and Decision 16

CHAPTER 3   ELEMENTS OF LOW-COST JIGS AND FIXTURES 21
Principles of Low-Cost Tooling..................................... 21
Reducing Jig and Fixture Costs 22

CHAPTER 4   METHODS OF LOW-COST DESIGN............................. 41
Designing a Low-Cost Tool 41
Tool Drawing Methods.. 42

CHAPTER 5   MULTIPURPOSE JIGS AND FIXTURES 47
Multipurpose Tool Applications.................................... 47
Multipurpose Tool Cost Estimating 48
Designing Multipurpose Tools...................................... 50
Additional Advantages of Multipurpose Tools 56

CHAPTER 6   UNIVERSAL TOOLING SYSTEMS 59
Designing Erector-Set Tooling...................................... 59
Constructing Erector-Set Tooling.................................. 60
Recording Erector-Set Tooling 64
Advantages of Erector-Set Tooling................................ 64

CHAPTER 7   FORMED-SECTION TOOL CONSTRUCTION 67
Commercial Premachined Sections 67
Structural Form Material 71

CHAPTER 8   MAGNESIUM TOOL CONSTRUCTION 75
               Characteristics of Magnesium 75
               Magnesium Fire Hazard 79
               Magnesium Stock for Tooling 80

CHAPTER 9   WOOD TOOL CONSTRUCTION 91
               Advantages and Disadvantages of Wood for Tooling 91
               Characteristics of Wood 92
               Processed Wood 96
               Designing and Fabricating Wood Tools 99
               Economics of Wood Tools 108

CHAPTER 10   EPOXY PLASTIC TOOL CONSTRUCTION 113
               Plastics 113
               Advantages and Limitations of Plastics 115
               Designing Epoxy Tools 118
               Fabricating Epoxy Tools 120

ADDITIONAL READINGS 131

INDEX 133

figures

Figure 2–1   Three hypothetical relationships of tool cost
to labor cost **10**

Figure 2–2   Graphic tooling analysis of three machining methods **13**

Figure 2–3   Typical machine setup for a lathe **14**

Figure 2–4   Master part method of transferring hole locations **15**

Figure 2–5   Master part used to reduce machine tool setup time **15**

Figure 2–6   Proposed tool evaluation form **17**

Figure 3–1   Method of specifying surface finishes on low-volume
tooling drawings **24**

Figure 3–2   Milling fixture constructed from prefinished
material **25**

Figure 3–3   Screw heads substituted for jig feet **26**

Figure 3–4   Conventional method of attaching keys to fixtures **27**

Figure 3–5   Simplified method of attaching keys to small fixtures **27**

Figure 3–6   Patented reamed-hole key **28**

Figure 3–7   Two methods of jig design **30**

Figure 3–8   Template jig constructed from prefinished material **31**

Figure 3–9   Template jigs **32**

Figure 3–10 Plate jigs **33**

Figure 3–11 Standard universal jig **34**

Figure 3–12 Universal jig ready for production **35**

Figure 3–13 Standard universal rotary table **35**

Figure 3–14 Limited-production milling application of a
rotary table **36**

Figure 3–15 Swivel vise **36**

Figure 3–16 Three-way vise **37**

Figure 3–17 Mill vise **37**

Figure 3–18 Vise converted to drill jig with bushing plate
and adjustable stop **38**

Figure 3–19 Vise converted to drill jig with adjustable
bushing plate **39**

Figure 4–1   Methods of drawing screw and dowel locations **42**

Figure 4–2   Simplified method of drawing hole locations **43**

Figure 4–3   Methods of drawing commercial parts **43**

Figure 4–4   Milling fixture drawn with conventional drafting
methods **44**

Figure 4–5   Milling fixture drawn with simplified, economical
                 drafting methods 45
Figure 5–1   Motor bracket made of stainless steel structural angle........ 50
Figure 5–2   Manufacturing process sheet for motor bracket of
                 Fig. 5–1... 51
Figure 5–3   Multipurpose fixture designed for motor bracket
                 of Fig. 5–1... 52
Figure 5–4   Setup for milling flats of the motor bracket 53
Figure 5–5   Setup for milling slots of the motor bracket.................... 54
Figure 5–6   Multipurpose fixture converted into a drill jig................. 55
Figure 5–7   Drawing for three steel links....................................... 55
Figure 5–8   Multipurpose drill jig for drilling the three links of
                 Fig. 5–7... 56
Figure 6–1   Typical process operation drawing for erector-set
                 jig assembly.. 60
Figure 6–2   Typical universal tooling system for both jigs and
                 fixtures.. 61
Figure 6–3   Typical erector-set tooling kit for fixtures....................... 62
Figure 6–4   Erector-set fixture 63
Figure 6–5   Template used in the construction of erector-set
                 tooling 63
Figure 6–6   Fixture being assembled from an erector-set tooling
                 kit 64
Figure 7–1   Standard commercial sectional shapes used for
                 tooling 67
Figure 7–2   Drill jig made from standard T-section tooling
                 material.. 68
Figure 7–3   Milling fixture made from standard L-section tooling
                 material.. 69
Figure 7–4   Drill jig made from standard U-section tooling
                 material.. 69
Figure 7–5   Milling fixture made from flat tooling plate 70
Figure 7–6   Grinding fixture made from T-section tooling
                 material.. 70
Figure 7–7   Drill jig made from standard structural form material....... 71
Figure 7–8   Steel part requiring both milling and drilling.................. 72
Figure 7–9   Combination drill jig and milling fixture made of
                 structural material for part of Fig. 7–8......................... 72
Figure 8–1   Magnesium tumble jig 76
Figure 8–2   Magnesium vibration testing fixture.............................. 78
Figure 8–3   Magnesium checking fixture.. 82
Figure 8–4   Magnesium extrusions 89
Figure 8–5   Locating fixture made with square magnesium
                 tubing.. 89
Figure 9–1   Cross-section of a tree showing growth rings................... 93
Figure 9–2   Forces or loads in relation to the growth rings in
                 wood... 94

Figure 9–3   Direction of warpage of a drying board........................... 95
Figure 9–4   Board warpage in relation to position in the tree.............. 95
Figure 9–5   Laminated construction to offset warpage....................... 96
Figure 9–6   Methods of protecting exposed edges of metal-
                   clad plywood.. 98
Figure 9–7   Types of butt joint construction 100
Figure 9–8   Rabbet and dado joints 101
Figure 9–9   Mortise-and-tenon joints 102
Figure 9–10   Reinforced miter joints 103
Figure 9–11   Dovetail joint..103
Figure 9–12   Mechanically reinforced joints.....................................104
Figure 9–13   Standard commercial drill bushings for wood
                   tooling 105
Figure 9–14   Standard screws used as stops and locators for wood
                   tooling 105
Figure 9–15   Commercial socket nut for use with wood tooling............106
Figure 9–16   Wooden drill jig incorporating standard tooling
                   hardware..107
Figure 9–17   Large wooden plate jig 108
Figure 9–18   Wooden milling fixture designed to hold an
                   irregular casting 109
Figure 9–19   Welded bracket 110
Figure 9–20   Wooden drill jig designed for bracket of Fig. 9–19 111
Figure 10–1   Drawing of an epoxy drill jig.......................................119
Figure 10–2   Laminated construction cross-section 121
Figure 10–3   Laminated plastic drill jig and scribe template.................122
Figure 10–4   Typical surface-cast epoxy plastic form 123
Figure 10–5   Surface casting—pour method.....................................123
Figure 10–6   Surface casting—squash method...................................124
Figure 10–7   Surface casting—pressure-pot method...........................124
Figure 10–8   Mass casting..125
Figure 10–9   Epoxy mass-cast vacuum fixture...................................125
Figure 10–10   Steps in the preparation and application of epoxy
                   paste...126
Figure 10–11   Epoxy lathe jaws for holding an irregularly
                   shaped part...127
Figure 10–12   Methods of attaching metal parts to epoxy tooling............128
Figure 10–13   Commercial drill bushings for epoxy tooling....................128
Figure 10–14   Method of locating and bonding drill bushings in
                   cast epoxy..129
Figure 10–15   Method of locating and bonding drill bushings in
                   laminated epoxy 129

tables

Table I–1   Relationship of Lot Size to Costs 6
Table II–1   Cost Analysis of Three Machining Methods 12
Table III–1   Cost Breakdown of Patented Reamed-Hole Keys vs.
                 Conventional Keys for Fixtures 29
Table VIII–1   Properties of Magnesium and Three of Its
                 Alloys 75
Table III–2   Mill Standard Sizes of Magnesium Tooling
                 Plate 81
Table VIII–3   Flatness Tolerances of Magnesium Tooling
                 Plate 81
Table VIII–4   Tooling Plate Comparative Costs 82
Table VIII–5   Cost Comparison of Three Jig Materials 83
Table VIII–6   Cost Comparison of Aluminum, Magnesium, and
                 Mild Steel Fixtures 84
Table IX–1   Properties of Wood 92
Table X–1   General Physical Properties of Epoxy Resin
                 (Pure State) 114
Table X–2   Properties of Plastic Tooling Filler 115
Table X–3   Plastic Tooling Construction Methods 118

Jigs & Fixtures for Limited Production

chapter

1

introduction

This chapter introduces the book by offering appropriate definitions and explanations of the terms we are going to discuss. The first section defines exactly what we mean by jigs and fixtures, and the next section compares and contrasts the concepts of mass production and limited production. The following section contains an analysis of the ways tooling costs affect manufacturing, describes some of the techniques used in limited production, and points out why jigs and fixtures fill the needs of those techniques. The Introduction closes with a statement of the challenges that face tool engineers in effecting tooling economies for short-run manufacturing while still retaining the benefits of well-planned, well-built tooling.

Many excellent books and articles, some of which are listed in Additional Readings, have been written about the fundamentals of jig and fixture design and tooling, and they are recommended to the reader as background material to assist in understanding the information in this book. Here, however, we will examine jigs and fixtures in relation to a special area of manufacturing—that of limited production.

DEFINITION OF TERMS

The terms *jig* and *fixture* are often used interchangeably, but for the purposes of this book we will define them more exactly. Jigs and fixtures are devices used to hold the workpiece in the correct position relative to a tool or gage during machining, assembly, inspection, and other manufacturing operations. A *jig* is a workholding device with a built-in capability for guiding the tool during a manufacturing operation. A *fixture*, in most cases, does not have built-in tool guidance. Its primary purpose is to locate the work, hold the work securely, and establish a relationship between the work and a machine tool by being fastened to the machine. The term *tooling* as used in this book will apply only to these workholding devices.

JIGS AND FIXTURES IN MANUFACTURING

Jigs and fixtures provide a means of manufacturing any desired number of parts to desired tolerances because they set the relationship between the work and the machine tool. To do so they are equipped with means for:

3

1) Locating the work
2) Clamping the work
3) Supporting the work
4) Holding all the elements together in a rigid unit
5) Guiding the tool (in the case of jigs)
6) Fastening and positioning the unit on the machine tool (in the case of fixtures).

Because of these characteristics, jigs and fixtures furnish several distinct advantages in manufacturing, in that they:

1) Ensure the interchangeability and accuracy of parts manufactured
2) Minimize the possibility of human error
3) Permit the use of unskilled labor
4) Reduce manufacturing time
5) Allow production of repeat orders without retooling.

MASS PRODUCTION AND LIMITED PRODUCTION

The interchangeability of parts produced by use of jigs and fixtures has played a great part in the development of mass production – the manufacturing of items in large quantities. When the dimensions of each part are controlled and held to close tolerances for long periods of manufacturing time, great numbers of them can be produced with confidence that they will fit properly during final product assembly. When large numbers of parts are produced, a manufacturer is almost always well advised to use jigs and fixtures. In this situation the advantages are lower tooling costs, decreased production time, and lower labor costs.

The advantages of jigs and fixtures for mass production are seldom questioned, but a popular belief exists that short runs cannot support tooling costs. This simply is not so. In limited-production situations the use of jigs and fixtures can facilitate production operations and actually aid in reducing fabrication costs.

Limited production in this context is not defined in terms of production lot size alone, for what is considered low volume to one manufacturer may well be classed as mass production by another, and even two individual companies producing identical items may classify a given production quantity differently. For this reason, the term *limited production* is defined as the production of a quantity substantially less than that which a company, industry, or facility is normally prepared to handle. There are, of course, shops and companies that specialize in manufacturing small lots. Although this definition of limited production does not apply to them – since limited runs are their normal production volume – they can use the information in this book to lower their tooling costs and operate more profitably.

It is well worth considering some of the circumstances which have forced manufacturing companies, willingly or unwillingly, to shorten their production runs – to manufacture smaller lots than they are capable of producing with existing equipment, capital, and labor resources. These may be listed as follows:

1) *Consumer demand.*—Because consumers demand a variety of models, colors, options, etc., manufacturers have been forced to shorten runs. Many mass-production programs for previously standard products have been cancelled, in fact, because of this factor. In addition to the necessity of producing shorter runs to satisfy the demand for product variety, manufacturers obviously must shorten a run when total demand slackens, as, for example, during an economic slack period or when increased competition has reduced sales.

2) *Reduction of lead time.*—Lead time may need to be reduced for any one of a variety of reasons—to secure a favorable market position for a new product, to get a seasonal product out in time to secure maximum sales, to meet the bid requirements of contractors, etc. Limited production may often provide the answer to shortening lead time in these circumstances, but it is feasible only if costs can be controlled. Inexpensive tooling offers a way of controlling costs and enables the company to shorten lead time.

3) *Correction of trouble areas.*—Through the use of short runs it is possible to find and correct trouble areas in the production line or in the product itself before full production has begun. Short production runs are worth considering, for example, when difficulties are expected because of a new or complex design.

4) *Spare parts and sample parts.*—If the demand for spare parts continues after the original production run has been completed, short runs are often necessary. In addition, a plant may be called upon to produce limited numbers of sample parts for sales, advertising, and promotional purposes.

5) *Prototype and experimental models.*—Limited-production methods are sometimes used in the manufacture of prototype or experimental products, where the quantities needed are usually small, particularly if a long testing period is to precede full production or if testing might indicate major changes in the original design.

In the opinion of many manufacturers, low production quantities will not justify the cost of jigs and fixtures, but many plants today are engaged in short-run manufacturing and are operating profitably. The decision to adapt mass-production techniques to short-run operations is a difficult one, without doubt, but low-cost tooling can make the decision easier and the chances of success greater.

ASPECTS OF LIMITED PRODUCTION

Tooling for limited production differs considerably from tooling for mass production, but in each case tooling costs are a major consideration. As an example, consider the case of a company planning to produce an item for which the tooling is estimated to cost $500. As shown in Table I–1, when tooling cost is held constant at $500 the cost per unit decreases geometrically as the size of the production lot increases.

Looking at these cost figures, assume that the company desires to produce

Table I–1. Relationship of Lot Size to Costs.

Lot Size	Tooling Cost	Other Costs*	Total Cost	Unit Cost
100	$500	$350	$850	$8.50
200	500	700	1200	6.00
300	500	1050	1550	5.17
400	500	1400	1900	4.75
500	500	1750	2250	4.50
600	500	2100	2600	4.33
700	500	2450	2950	4.21
800	500	2800	3300	4.13
900	500	3150	3650	4.06
1000	500	3500	4000	4.00

*The assumption here is that other costs – direct labor, direct material, and overhead – increase in direct proportion to increases in the number of units manufactured. This is seldom the case in actual production situations.

a lot of 300 units but that the per unit cost may not exceed $4.50. With a few calculations we can quickly determine that to achieve this unit cost through tooling cost reductions we must somehow lower the cost of the necessary jigs and fixtures to $300. Thus the total production costs ($300 for tooling plus $1050 for other costs) will equal $1350, and this sum, when divided by 300 (the number of units in the shortened run) will give us the desired per unit cost of $4.50.

In most cases, limited production runs rely upon general-purpose machines, whereas large-volume manufacturing machines are highly specialized and inflexible. This factor, too, has an important effect on tooling costs. For example, the manufacturer of the 300-unit lot in the example above may decide to use general-purpose machine tools and jigs and fixtures which would cost him a total of $600 but would lower his other costs to $700. He would then be able to produce his lot at a per unit cost of $4.33. On the other hand, if the same manufacturer doubled his expenditure for specialized machine tools, with his costs for specialized labor increasing in almost direct proportion, his per unit costs would be lowered only after a much longer run.

We have seen here that costs can be reduced by using less expensive tooling or by using tooling that lowers costs in other areas. It is sometimes argued that jigs and fixtures should be eliminated entirely when small lots are produced. This may be true in some cases, but the advantages of workholding devices, as listed earlier, apply as well to limited-production operations. Interchangeability is improved, more accurate parts can be manufactured, and, most importantly, labor costs are reduced. While the use of special tooling is primarily intended for the last purpose, the advantages of improved interchangeability and the ability to supply replacement parts may often be equally significant. Savings will also appear in assembly costs because of improved interchangeability, and this reason, and the improved functioning of the part itself, should be seriously considered as sufficient cause for "tooling up" for low production.

Pilot Runs

Limited-production jigs and fixtures are becoming more popular for pilot or preproduction runs. This type of tooling is helpful in locating and correcting possible trouble areas in manufacturing before full-scale production commences. A limited amount of capital invested for preproduction runs does not cause much of an expense in the correction of manufacturing processes or tools yet provides valuable information for the design of the permanent production tools. In contrast, changes required for permanent tools or operations designed before the pilot-lot run would be very expensive.

Medium-Volume Runs

The principles of limited-production tooling can also be applied to medium-volume manufacturing. Although low-cost jigs and fixtures may wear out and have to be replaced in order to complete the production run, it is often cheaper and easier to replace them than it is to design and build a permanent tool. Low-cost workholding devices should certainly be considered when there is a possibility of product design changes midway in the production run which would require alterations in a more costly jig or fixture.

Single Part Production

Occasionally a jig or fixture is fabricated merely to produce a single part. One part would not normally justify special tooling, but if the part is to be made of expensive or scarce material, special tooling can minimize the possibility of waste through machining error. Obviously, tooling for this purpose must take full advantage of all possible cost-saving techniques.

Numerical Control and Tooling

Today's modern, precision, numerically controlled machines often make it possible to dispense with jigs and fixtures for limited production, but the high cost of these complex machines limits their use to some extent. In contrast, jigs and fixtures represent a smaller capital investment, and they nearly eliminate the need for skilled labor.

THE CHALLENGE TO TOOL ENGINEERS

The increase in short-run manufacturing has created a corresponding need for low-cost tooling. Low-cost jigs and fixtures can offer many of the same advantages in small-lot production that permanent tooling offers in mass production. Through the use of less expensive jigs and fixtures, a manufacturer can find it possible to manufacture and sell products in small quantities that previously would have offered no opportunity for profit.

Many companies have already recognized the advantages and opportunities available to them through the use of inexpensive tooling. But while these manufacturers have realized the need for economy, others have not. As a result, production lines of the latter group have frequently been overtooled, and their costs have climbed correspondingly. Dollars that should have shown up as profit on the income statement instead are listed under "cost of goods sold." Herein lies the challenge for tool and manufacturing engineers. It is their responsibility to recognize areas where less expensive tooling can be utilized. It is their responsibility to design and, in many cases, to build this tooling. The challenge then is theirs to see that lost dollars spent on excessive tooling are instead re-tained, and upon their success in meeting this challenge rests the answer to the question whether company profitability will be improved through the applica-tion of sound tooling policy.

chapter

2

tooling for
limited production

The last chapter defined jigs and fixtures, discussed their advantages and cost relationships, and pointed out some of the factors that manufacturers should consider in deciding whether to tool up for limited production. The chapter ended with a challenge to the tool engineer to find new and improved ways to lower the costs of low-volume production through the use of jigs and fixtures.

This chapter goes into more detail regarding the effects of tooling upon manufacturing costs and the relationship between labor and tooling. It examines the analyses tool engineers should make to provide management with facts for decision making, and also describes some of the choices that might be made in response to the need for low-cost tooling. The chapter closes with a discussion of the problems involved in evaluating proposals for limited-production jigs and fixtures and other low-cost production methods.

MANUFACTURING COSTS

Manufacturing costs can be classified as direct and indirect costs. *Direct costs* are those that are assigned directly to each part produced. *Indirect costs* are those that cannot be directly identified with the manufacture of a specific item. For the purposes of this book we will confine our attention to direct costs.

Direct costs include the following expenses:
1) *Investment charges on tooling.* — The actual costs incurred for tooling.
2) *Direct labor costs.* — Salary and fringe benefits for personnel engaged directly in production operations.
3) *The cost of material.* — Expenses for material that becomes part of the finished part or product. As mentioned in Chapter 1, the use of costly material can increase tooling expenses. Such material is not generally used, however, and since material is usually chosen before tooling decisions are made, material costs do not concern us here.

TOOL COSTS VS. LABOR COSTS

Two facts must be kept in mind when tooling up for limited production. First, labor costs can be reduced by using jigs and fixtures, and second, conversely, undertooling causes labor costs to increase. This relationship is demonstrated in Fig. 2–1. Note that variable tooling and labor costs make up the total

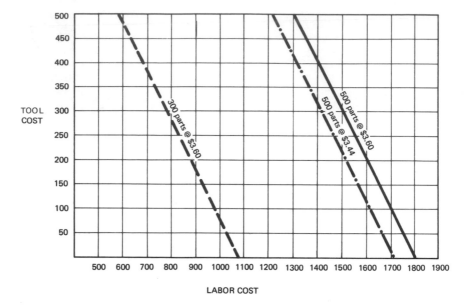

Fig. 2-1. Three hypothetical relationships of tool cost to labor cost.

cost when the total amount that can be spent is known. For example, if 500 parts are to be produced at a cost of $3.60 per unit and a total of $1800, then $1200 can be spent for tooling and $600 for labor, or alternatively $500 can be spent for tooling and $1300 for labor. These are, or course, only two of the many possible sets of complementary relationships.

It is important to know exactly how much can be spent on tooling for limited production runs. Three factors will affect the decision regarding the total expenditure:

1) Size of the production run
2) Tooling cost variations
3) Labor skill levels.

Each of these is discussed below.

Size of Production Run

The permissible investment level for tooling depends largely upon the number of parts to be produced—both initially and in total. This relationship of tooling cost to production run size can also be seen in Fig. 2-1. Assume that a manufacturer plans to produce 300 fittings at a maximum per unit cost of $3.60. The total cost he is willing to incur is then $1080. Assuming a labor rate of $8.00 per hour, the production planner must utilize tooling that will permit production of at least 2.2 parts per hour in order to cover labor costs ($8.00/hr ÷ $3.60/part). An additional reduction in production time must be made to cover tool costs. If the manufacturer uses tools that will produce 2.5 parts per hour and will cost only $120, he will then be able to produce his lot of 300 fittings within the limit of $1080.

If the manufacturer later decides to produce an additional lot of 200, labor costs on the second run will be $640, bringing total costs to $1720 but reducing per unit costs for both runs combined to $3.44. If he must replace tooling to complete the run, however, total costs will reach $1840, and unit cost will be $3.68.

Variations in Tool Costs

Labor costs are usually incurred on an hourly basis. Because of this, an increase in production rate has the effect of lowering the labor cost per unit. Although the cost of jigs and fixtures is not always related to their quality, the rule is generally true that a more expensive tool enables faster parts production than a cheaper tool and thereby saves money on labor. If a more expensive tool does not produce parts faster than a cheaper tool, it is not worth the extra cost unless improved accuracy or some other advantage is desired. On the other hand, cheap tools do not lower costs if they slow operations excessively.

Labor Skill Levels

As mentioned in Chapter 1, one of the main advantages of tooling is that it reduces the level of worker skill needed over a wide range of operations. Tooling permits labor specialization and allows the training and use of less skilled operators without jeopardizing finished product quality. In effect, jigs and fixtures transfer the needed skill from the operator to the tool while at the same time reducing the chances of human error. Obviously the use of lower-skilled workers reduces labor costs.

PRESENTING THE TOOLING ESTIMATE

Management often looks upon jigs and fixtures as costly, but necessary, evils, and the tool engineer may sometimes feel frustrated by what appear to be delays on the part of management to accept his proposals. He must remember, however, that management's responsibility is to consider each tooling problem in the light of potential profit or loss. For this reason the tool engineer should present his tooling estimate in terms of cost savings and profit potential, and he must be prepared to justify proposed expenditures. The tooling estimate must include at least these three points:

1) The cost of the proposed jig or fixture, including design costs
2) The estimated hourly production rate using the proposed jig or fixture
3) The projected profit or savings if the jig or fixture is used.

When the cost and the hourly production rate of a jig or fixture have been estimated, the profit or savings it will create can be calculated. Although there are several methods and formulas for determining whether or not the expenditure is justifiable, the following formulas are recommended because of their relative simplicity:

$$B = \frac{E}{C - D} \tag{1}$$

$$F = A(C - D) - E \tag{2}$$

Where:
- A = Total quantity of parts to be manufactured
- B = Minimum quantity of parts necessary to justify tools
- C = Estimated cost of producing parts (per unit) with available tools
- D = Estimated cost of producing parts (per unit) with special tools
- E = Estimated cost of special tools (design and fabrication)
- F = Possible savings through use of special tools

Example: 275 castings are to be drilled. Estimated drilling cost without special tools is $.80 each per casting. Drilling the castings with a specially designed jig is estimated at $.20 each. The jig will cost approximately $125 to design and build. What will be the minimum quantity of parts necessary to justify tooling?

$$B = \frac{E}{C - D} \tag{1}$$

$$= \frac{\$125}{\$.80 - \$.20}$$

$$= \frac{\$125}{\$.60}$$

$$= 208 \qquad \text{Minimum number of castings necessary to justify tooling.}$$

What will be the savings?

$$F = A(C - D) - E \tag{2}$$

$$= 275(\$.80 - \$.20) - \$125$$

$$= 275(\$.60) - \$125$$

$$= \$165 - \$125$$

$$= \$40 \qquad \text{Savings through the use of the special jig.}$$

It is often necessary for the tool engineer to determine the most economical approach to manufacturing a given quantity of parts by trying several methods and tools. He might make an analysis, for instance, of machining 300 fittings by use of three known methods as shown in Table II–1.

Table II–1. Cost Analysis of Three Machining Methods.

Method	Time (min/unit)	Labor Cost*	Tool Cost	Total Cost (Labor and Tooling)
1) Without Special Tools	10	$400	——	$400
2) With Fixture (Simplified)	5	$200	$115	$315
3) With Fixture (Conventional)	4	$120	$270	$390

*Labor rate = $8.00/hr.

This information can then be transferred to a graph to determine when each method becomes most economical. The graph shown in Fig. 2–2 indicates that machining the fittings without special tools (Method 1) would be economical for lots of fewer than 180 pieces. For larger lots, the use of a simplified, specially designed fixture (Method 2) would be more economical. But Method 3, though it would offer greater productivity potential, would not be economically feasible for just the 300 fittings. Obviously, Method 2 is the correct choice.

When tool cost exceeds an established minimum, the tool engineer should prepare a mathematical and graphic analysis explaining the reasons for his selection. Minimum cost should be established by management and will vary from one company to another.

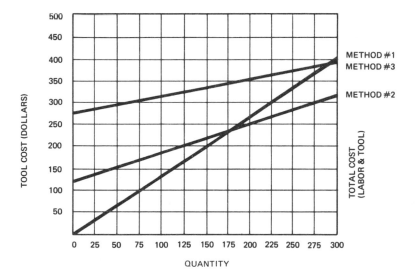

Fig. 2–2. Graphic tooling analysis of three machining methods.

The engineer's economic analysis of proposed tooling is more important in planning a limited-production manufacturing program than an automated or mass-production program. Tooling expenditures for the latter are more readily absorbed. The manufacturer's general rule of thumb when selecting tools or processes for a limited-production program is that every tool should pay for itself with the initial production quantity for which it was intended. Any exception to this rule should be considered carefully and approved only by executive decision.

MANUFACTURING WITHOUT SPECIAL TOOLING

By preparing an economic analysis, the tool engineer may find operations or parts that do not warrant special tools. There will always be a minimum production volume below which the cost of tooling cannot be justified. Nevertheless, improved accuracy, interchangeability, etc., may often warrant adoption of an otherwise questionable tool, and only after complete investigation should an operation or part be handled without jigs or fixtures.

Even when special tooling cannot be justified, another low-cost production method can usually be found. Finding the method merely requires greater ingenuity on the part of the tool or process engineer. Three methods that might be used are:

1) The machine setup system
2) The master part system
3) The limited-production group system.

Machine Setup System

Many major contributions to successful low-volume operations can be credited to ingenious money-saving techniques and setups using conventional tools and equipment. Usually the setup arrangements are recorded in the form of sketches and are attached to process or route sheets. Sketches should be provided for all operations that lack special tooling in order to avoid confusion in the shop. A typical sketch illustrating a lathe setup is shown in Fig. 2–3.

Master Part System

If an item does not warrant special tooling, there may be an economic advantage in using the master part system. The system, as the name implies, uses

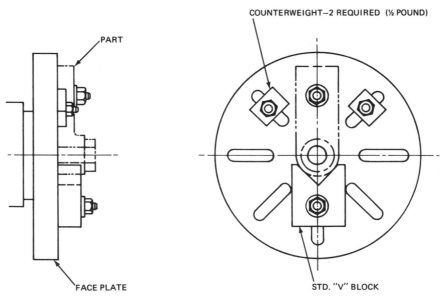

MACHINE #1231 14″ LATHE

OPERATION 10 – TURN .750 DIAMETER AND
BORE .500 DIAMETER HOLE

Fig. 2–3. Typical machine setup sketch for a lathe.

a simulated or actual part fabricated to slightly closer tolerances than the intended product and used as a template or jig.

When a master part is used as a jig for locating holes, the master and the part to be worked are clamped together as shown in Fig. 2–4, and the hole locations are transferred from the master to the part with a duplicating punch. If possible, hardened bushings are inserted into the master part, making it possible

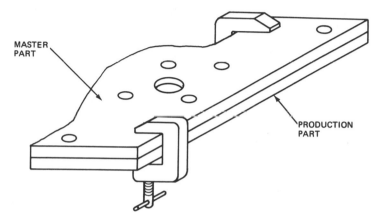

MASTER
PART

PRODUCTION
PART

Fig. 2–4. Master part method of transferring hole locations.

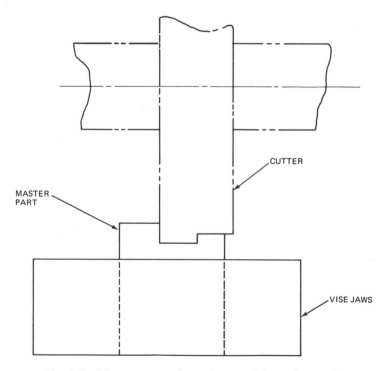

CUTTER

MASTER
PART

VISE JAWS

Fig. 2–5. Master part used to reduce machine tool setup time.

to transfer holes directly by drilling. The addition of drill bushings improves accuracy and increases tool life.

Master parts are also used to reduce machine setup time. The master part is simply placed into its holding device, and the machine's spindle or cutter is "zeroed" to the corresponding master surface as shown in Fig. 2–5. The master is then removed and replaced with an actual part. This setup method is very efficient and eliminates much trial and error.

Inspection of finished or in-process parts can also be simplified by the use of a master part. The master is used as a template together with a surface plate for the purpose of setting indicators. This method eliminates the calculations usually necessary when gage blocks are used.

The master part system of manufacturing is one of the simplest methods of producing interchangeable parts, but it is limited to parts of simple configuration and to very small quantities. It is seldom used for the manufacture of parts which require great accuracy.

Limited-Production Group System

When it is impractical to employ the methods discussed above, it is necessary to manufacture with available standard tools and equipment. Manufacturing without a completely tooled program, however, requires highly skilled operators, and so a nucleus of qualified, highly trained personnel should be organized specifically for the purpose. The selected operators should be given responsibility for processing—working directly from part prints—and for determining the type of machines and tools necessary in each case. Since they may be required to work in all stages of production—from raw material to finished product—they must be capable of operating most types of conventional equipment as well as doing their own in-process inspection.

Such a limited-production manufacturing group can often be successfully organized within an active, high-production department or, with less capital investment, combined with a tool fabricating section. Either of these two methods of operation, of course, requires careful scheduling so that regular production programs are not interrupted.

For greater efficiency, the group should be completely divorced from the main production activities. This separation of activities may require a new department equipped with the most versatile tools available and with limited-quantity or prototype manufacturing as its specific purpose.

Many companies are realizing an additional benefit from such departments, for they are able to manufacture spares in small quantities as needed. This side benefit provides the capability of satisfying customer demands for fast delivery of service parts while maintaining stock inventories at economic levels.

EVALUATION AND DECISION

After the tool engineer has obtained all the necessary facts, has evaluated his information, and has come to a conclusion whether a given operation justifies tooling, he must advise management in its decision whether to use jigs and fixtures or some alternative in limited production.

The decision whether tooling is justified in a particular situation has to be based on adequate data, and the data should be presented in an orderly fashion. One way for the tool engineer to present his information is on a "proposed tool evaluation form." Such a form should be concise and should contain only the information required to make sound decisions. A suggested form is shown in Fig. 2–6. If more detailed information seems desirable, mathematical

---

PROPOSED TOOL EVALUATION

PART NAME AND NUMBER: *HOUSING – BL-38621*

QUANTITY: *50 Pcs.*

OPERATION: *DRILL (4) HOLES*

TOOL: *DRILL JIG*

ESTIMATED COST: *$210.00*

ESTIMATED SAVING: *NONE*

SUGGESTED ALTERNATE METHOD: *MANUFACTURE IN LIMITED PRODUCTION DEPARTMENT*

STATE REASON FOR TOOL REQUEST (If No Saving):

1) EXCELLENT POSSIBILITY OF RE-ORDER
2) CLOSE TOLERANCES MAY BE DIFFICULT TO MAINTAIN WITHOUT JIG.

PREPARED BY: *J. Loomis*   DATE: *1-14-65*

MANAGEMENT DECISION:

APPROVED: ☐   DISAPPROVED: ☑

COMMENTS: *ROUTE JOB TO LIMITED PRODUCTION DEPARTMENT*

AUTHORIZED SIGNATURE: *C. Daigle*

Fig. 2–6. Proposed tool evaluation form.

and graphic tooling analysis data may also be included as an attachment to the form.

The question that management must answer falls into two closely related parts:

1) Should a new tool be procured?
2) If not, which of the alternative manufacturing methods will do the job most effectively?

These questions should be answered by the tool engineer before the evaluation form is presented to management, and the engineer should therefore develop the habit of using all the data he has collected to compare and evaluate the overall cost of the new tool and each of the alternate methods. The only logical means of rendering a decision is based on that data.

This does not mean that every tool evaluation presented should be overloaded with facts, for different tooling problems require different amounts of information. In case the proposed tooling does not involve much capital, detailed information may not be necessary. In contrast, if the tooling will require a great expenditure, detailed and accurate information is required. In any event it is up to the tool engineer to evaluate the data he has developed and to be as thorough in his presentation as the facts permit and the problem requires.

When a decision regarding tooling must be made, it is often important to know when to recommend or approve a tool that cannot be justified in dollars saved. A jig or fixture can generally be justified for one of the following reasons:

1) It will provide a substantial savings
2) It is considered a necessity because of one of its inherent advantages.

Jigs and fixtures which permit immediate cost savings and those whose total cost can be recovered during the initial production run are ideal. The tool engineer recommending the use of tooling in this category will normally meet with no objection from management. The justification for the use of such tooling is obvious. A recommendation to purchase or build tooling that does not offer immediate savings, or which requires more than one production lot for full cost recovery, is less likely to be approved. The use of this type of tooling is justifiable on occasion, but only when one of the inherent advantages of jigs and fixtures is so significant as to override immediate, short-run cost considerations.

Justification of jigs and fixtures for low-volume production, especially when there are no savings on tool cost involved, is difficult. It is necessary to weigh the inherent advantages of tooling against the required cash outlay. The difficulty of the evaluative process is compounded by the fact that tangible factors (cash) must be set off against intangible factors (advantages). There is no common denominator in this problem, and the tool engineer may be strongly tempted to resort to guesswork. To do so, however, is to leave the most important tooling decisions to chance.

The only way out is to translate the advantages of tooling into financial terms. One advantage is reduced manufacturing time. Then how much operator time is saved? What is the labor cost per hour? Another advantage is improved interchangeability and accuracy. Then how much are rejects and rework reduced? What is this worth in dollars? After each tooling advantage is analyzed in financial terms, the individual sums are added. By comparing the resulting total with the required cash outlay for the contemplated tooling, a more realistic decision can be made. Obviously, not every advantage can be expressed in quantitative terms, and of course the translation of tooling advantages into dollar and cent terms involves estimates and is to some extent subjective. Nevertheless, rather accurate results are possible. The main advantage of this approach is that it provides an orderly method of considering all relevant factors.

The decision whether to provide a jig or fixture for limited production usually means the difference between profit and loss. Only by summing up all the facts and problems, evaluating all the compiled data, and using sound judgment in decision making can a limited-production manufacturing program be made to pay off.

chapter
3
elements of
low-cost jigs & fixtures

The tool engineer is responsibile for finding ways to reduce tooling costs to the absolute minimum. His job is doubly difficult and most vital in limited production, for as shown in the previous chapters, every aspect of tooling must be weighed carefully to find ways to lower tooling costs and make small-lot production profitable.

This chapter discusses the principles of low-cost tool design that must guide the tool engineer if he is to meet the demands of his job in limited-production operations, and it describes techniques and materials that he may find useful in following those principles.

PRINCIPLES OF LOW-COST TOOLING

Many of the usual principles of tool design cannot be followed when tooling costs are reduced, but through careful analysis of the product's design the tool engineer can find ways to make jigs and fixtures that are low in cost but which fulfill basic requirements—locating, clamping, and supporting the work; holding all elements rigidly together; and guiding the tool or fastening the unit to the machine tool.

There are certain principles of design, however, which low-volume manufacturing will especially warrant. These principles are as follows:

1) *Reduction of machining.*—Low-cost jigs and fixtures should be simple and require a minimum of machining or grinding. There are several techniques the tool engineer may use to avoid unneeded machining, and they are discussed later in this chapter.

2) *Use of wider tolerances.*—Unnecessarily close tolerances are perhaps the greatest single factor contributing to excessive jig or fixture cost. Tolerances should always be examined to determine their practical need in each case. This principle is also examined further.

3) *Use of basic and standard designs.*—There are certain basic types of jigs and fixtures or standard components that are suitable for low-volume production and that are also less costly than others. The tool engineer should examine these tools, listed in the last section of this chapter, for application to his problem.

4) *Elimination of rapid loading and unloading features.*—Because a high production rate is not as important over a short run as over a longer run, the tool engineer might well consider the elimination of rapid loading and

unloading features which do not necessarily have to be incorporated into the tools.

5) *Reduction of heat treatment.* — In limited production, wear will seldom create problems, so heat treatment of wear surfaces is generally unnecessary and may often be eliminated.

The flexibility in tool design allowed by these principles has made the use of many new tool design techniques and tool construction materials possible. In the past, most jigs and fixtures were made of steel, and seldom was thought given to any other construction material. Today, however, new materials and fabrication techniques are constantly being developed, and investigation should prove many of them suited to the tool engineer's needs.

Materials such as steel and magnesium plate, extruded forms, and wood and plastics are replacing more costly jig and fixture material. Coupled with the broad range of standard "off-the-shelf" tooling components, such materials can provide substantial savings. Further savings are possible through the use of universal-type jigs and fixtures and universal "erector-set" tooling systems, which are becoming increasingly popular.

REDUCING JIG AND FIXTURE COSTS

Jigs and fixtures themselves can be thought of as low-volume products. The direct cost of any jig or fixture includes the direct costs of the tools, labor, and material used in its fabrication. Because tools are produced in limited quantities, it is often difficult to reduce the costs involved in making them, but these costs *must* be reduced in proportion to the tools' own smaller production quantities. This cost reduction can usually be accomplished by eliminating all unnecessary costly features, some of which may appear ingenious or important without actually contributing much to functional requirements. (Of course, the tool engineer's aim is to design the simplest tool that will do the job, but if complex and costly tooling is actually necessary it should not be discouraged.)

When analyzing the design of a proposed limited-production jig or fixture, the tool engineer must be alert to recognize costly features and offer possible substitutions that are less expensive. The first three principles of low-cost design are presented below, with a discussion of each that includes cost-saving techniques and substitutes that may be of value in analyzing or designing a limited-production jig or fixture.

Reduction of Machining

Every machining cut that a toolmaker performs adds to tool cost. Therefore, the machining of surfaces that have no relation to the tool's accuracy or function should be eliminated, and all surfaces that do not require a fine machine finish should be specified on the tool drawing. In most cases it is only necessary to specify the surfaces that do not require finishing, especially those that might otherwise receive better finishes than are economically practical.

Surface finishes in tool drawings can be expressed in terms of microinches or with symbols and notes. Since not everyone is familiar with the microinch system, a conversion chart should be used with it. For this reason, the symbol and note system is preferable to avoid confusion.

The symbol and note method is illustrated by the drawing of the limited-production drill jig shown in Fig. 3–1. Note that 1.0-in stock is specified for the base (Detail 2) with no allowance for grinding. The dimensions of the base after grinding would, of course, have been slightly less, but exact size was not considered important. If the notation STOCK had been omitted from the thickness dimension, the toolmaker would have fabricated the base from the next larger size material and finish machined it to exactly 1.0 in, with resulting waste in material and machining time.

Dimensions appearing with the word STOCK and without finish or grind marks indicate that the material finish as received from the steel mill is acceptable. The words SAW CUT (see Fig. 3–1) indicate that the stock is to be sawed to specifications without additional machine finishing. Jigs and fixtures left unfinished will not have a first-class appearance, but their functionality will not be adversely affected. The objective is to reduce costs while retaining functionality.

Prefinished Materials. Machining costs can be further reduced by the use of commercial prefinished materials. An example of a low-production milling fixture constructed completely from prefinished stock is shown in Fig. 3–2. As the drawing indicates, fabrication of the fixture consists simply of cutting the details from bar stock to the lengths shown, then drilling, tapping, and reaming for screws and dowel pins. The estimated fabrication time for the fixture is three hours — an obvious saving over conventional methods of constructing tooling.

Jigs and fixtures constructed of prefinished materials are usually fastened together mechanically. They should not be welded because possible heat distortion would require a machining or straightening operation and thus defeat the purpose of using prefinished stock.

Tooling plate, drill rod, and cold-rolled steel are three examples of prefinished materials that can be used for jigs and fixtures with a minimum of machining.

Tooling Plate. Material such as tooling plate, available in many standard sizes and ready-ground to close tolerances, is being used for limited-production tools with considerable success. It generally requires little or no additional machining and can be hardened if necessary.

Drill Rod. Drill rod is a carbon tool steel that is cold drawn and usually ground or polished to accurate limits. It is supplied in rounds of all wire gages up to and including one inch in diameter and usually in three-foot lengths. It has found wide use in low-cost tool construction.

Cold-Rolled Steel. Cold-rolled steel is valuable to the tool engineer; it is inexpensive, has a very smooth finish from the rolling process, and is available in a number of shapes and sizes. Although it is not manufactured to the accuracy of tooling plate or drill rod, mill tolerances are usually adequate without further machining. Like tooling plate, cold-rolled stock can be case hardened when necessary.

Screw Heads. Many other means of reducing machining operations should be considered when limited-production tooling is analyzed and designed.

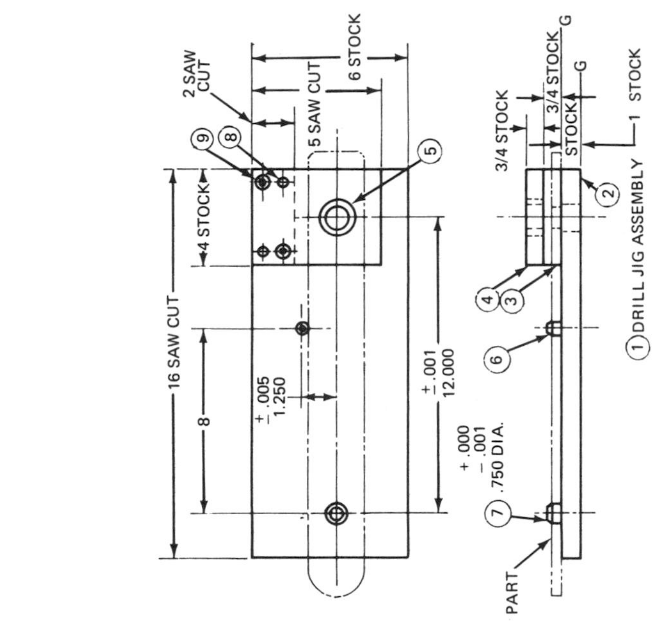

DET.	DESCRIPTION	REQ.	MAT'L	FIN. STOCK SIZE
9	SOC. HD. CAP SCR.	2	STD.	5/16 - 18 x 2 LG.
8	DOWEL PIN	2	STD.	1/4 DIA. x 2 LG.
7	LOC. PIN	1	D. R.	.750 DIA x 1 3/8 LG.
6	DOWEL PIN	1	STD.	5/16 DIA. x 1 LG.
5	DRILL BUSHING	1	STD.	1/2 I. D. x 3/4 LG.
4	BUSHING PLATE	1	C. R. S.	3/4 x 4 x 5
3	SPACER	1	C. R. S.	3/4 x 2 x 4
2	BASE	1	M. S.	1 x 6 x 16
1	ASS'Y	1		

DRILL JIG

DRAWN BY – C. D. D.
DATE – 1 - 14 - 65

APPROVED – F. S.
DATE – 1 - 18 - 65

DRAWING NO.

N – 2232

Fig. 3–1. Method of specifying surface finishes on low-volume tooling drawings.

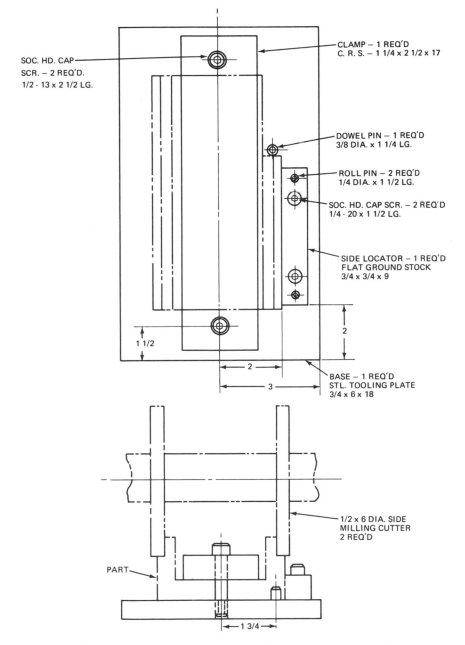

SOC. HD. CAP
SCR. – 2 REQ'D.
1/2 - 13 x 2 1/2 LG.

CLAMP – 1 REQ'D
C. R. S. – 1 1/4 x 2 1/2 x 17

DOWEL PIN – 1 REQ'D
3/8 DIA. x 1 1/4 LG.

ROLL PIN – 2 REQ'D
1/4 DIA. x 1 1/2 LG.

SOC. HD. CAP SCR. – 2 REQ'D
1/4 - 20 x 1 1/2 LG.

SIDE LOCATOR – 1 REQ'D
FLAT GROUND STOCK
3/4 x 3/4 x 9

2

1 1/2

2

3

BASE – 1 REQ'D
STL. TOOLING PLATE
3/4 x 6 x 18

1/2 x 6 DIA. SIDE
MILLING CUTTER
2 REQ'D

PART

1 3/4

Fig. 3–2. Milling fixture constructed from prefinished material.

For example, fillister or socket-head cap screws should not be recessed. Recessing (or counterboring) is not an expensive operation, but when considerable numbers of screws are used, the operation cost will be multiplied. The protruding screw heads can also serve as jig feet or rest buttons as shown in Fig. 3–3. To ensure alignment and uniformity, the screw heads may be machined or

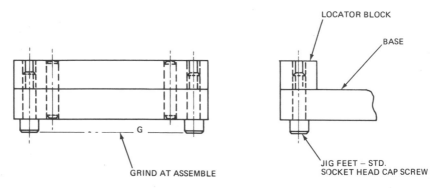

Fig. 3–3. Screw heads substituted for jig feet.

ground after assembly of the tool. When a jig or fixture must be turned over — as in the case of a tumble jig — and buttons or feet are required on several sides, the tool can usually be designed to incorporate assembly screws on those sides. The substitution of screw heads for jig feet or rest buttons eliminates not only the cost of special components but also the expense of installing them.

Groove or Spring Pins. It is usually necessary for the details of a jig or fixture to be accurately located, and this requirement calls for the use of at least two dowel pins per detail. Dowel pins, however, need accurately reamed or lapped holes, and this process involves extra expense. To reduce the amount of machining, grooved or spring pins should be substituted for dowel pins when possible so that only a drilled hole is needed. However, grooved or spring pins should not be used when alignment or location tolerances are less than .001 in.

Fixture Keys. Milling fixtures often require keys attached to their bases to fit in the locating slots of milling-machine tables. The keys provide a definite orientation and alignment of the fixture on the table in relation to the arbor and the cutters. The keys are usually set into a machined slot or grooved in the fixture base and are secured by screws as illustrated in Fig. 3–4. This method is expensive because the slot must be accurately machined or ground and the holes must be drilled and tapped for the screws.

Fig. 3–5 shows another, less expensive method that is ideal for small fixtures. A reamed hole, with a diameter matched to the keyway or T slot of the machine table, is machined into the fixture base. A dowel or stud is then press-fitted into the hole.

The patented, reamed-hole key shown in Fig. 3–6 is also available. It is specifically designed to eliminate the milling of keyways in fixtures. The key is a self-contained unit that can be mounted in either a through or blind hole. In addition, it provides interchangeability between fixtures and machines since all key sizes fit the same hole diameter. This means that various sizes of keys

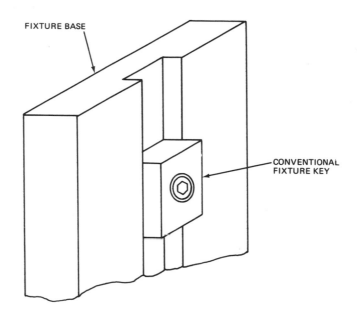

Fig. 3–4. Conventional method of attaching keys to fixtures.

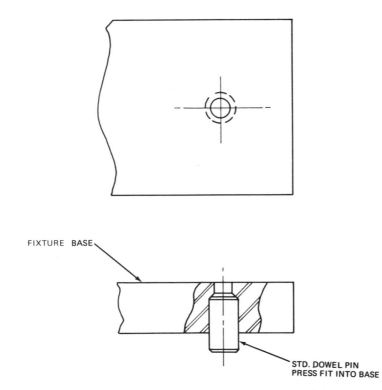

Fig. 3–5. Simplified method of attaching keys to small fixtures.

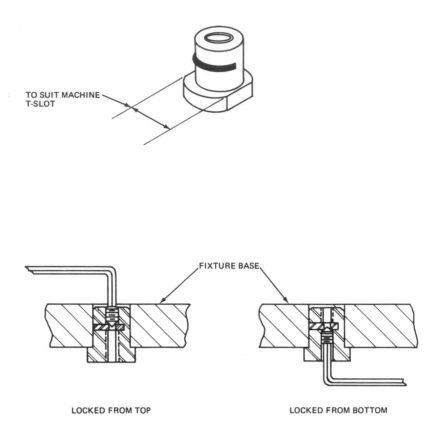

TO SUIT MACHINE
T-SLOT

FIXTURE BASE

LOCKED FROM TOP   LOCKED FROM BOTTOM

Fig. 3–6. Patented reamed-hole key. (*Courtesy, Standard Parts Company*)

can be adapted to the fixture to suit the T slot of any machine — with consequent time savings.

Savings in dollars and cents with the patented key can amount to more than 50 percent. In a cost comparison by its manufacturer, as detailed in Table III–1, installation of a set of conventional keys onto a fixture was estimated to be $19.62, including labor and material. The reamed-hole keys, on the other hand, cost only $9.70 — saving $9.92 per fixture.

Standard Components. Before designing or fabricating jig or fixture components such as clamps, locators, rest buttons, etc., it is wise to investigate the use of standard off-the-shelf items. It is usually more economical to buy commercial components than to fabricate similar parts. Also, initial savings may often be increased, because standard tooling components can frequently be modified to suit a design that would otherwise need specially fabricated components.

The possibilities of reducing tooling expense through the use of standard, commercially available tooling components are many and varied, and it is to the tool engineer's advantage to keep himself informed of the types of components available and of current developments in this field.

**Table III-1. Cost Breakdown of Patented Reamed-Hole Keys vs
Conventional Keys for Fixtures.**

Conventional Key Method	
Two milled keyways at prevailing shop rates	$15.00
Drill and tap for two .25-in screws	3.00
Two standard fixture keys @$.75	1.50
Two .25-in cap screws @$.06	.12
Cost Complete	$19.62
Patented Reamed-Hole Key Method	
Drill and ream while on jig borer	
Two .625-in holes	$4.00
Two .6875-in patented reamed-hole keys @$2.85	5.70
Cost Complete	$9.70
Conventional Key Method Total	$19.62
Patented Reamed-Hole Key Method Total	9.70
Savings	$9.92

(Courtesy, Standard Parts Company)

Use of Wider Tolerances

The tolerance specifications for a jig or fixture depend on the accuracy requirements of the product or the operations it will perform. The accuracy called for will directly affect tooling cost. Close tolerance specifications on tool drawings increase costs considerably. For this reason it pays the tool engineer to review all tool drawings and to relax tolerances wherever possible. In general, tool tolerances directly related to the product should not be less than 40 percent of corresponding product tolerances. It costs approximately 30 percent more to machine to tolerances of .0005 in. than to .001 in, and when tolerances are reduced from .001 in. to .00025 in, machining costs can increase up to 50 percent.

Jigs or fixtures can be designed without great accuracy yet still produce favorable results. This fact is illustrated by the two drill jigs shown in Fig. 3–7. Designed for the same operation, they are basically identical; their only difference lies in the locators or stops.

The jig shown in Fig. 3–7a orients the part in relation to the drill bushing by means of three rest buttons, so the relationship must be accurate. In contrast, the jig illustrated in Fig. 3–7b will produce the same product, but the bushing need not be located accurately because the three locators or stop screws can be adjusted to control the hole location in the part. Once set, the adjusting screws can be locked with a jam nut and sealed to prevent tampering or accidental changes in adjustment.

Both designs were submitted for competitive bidding. The cost quoted for the adjustable jig in Fig. 3–7b was $60.00 as compared with $85.00 for the jig in Fig. 3–7a.

The use of adjustable locators can reduce the cost of a jig or fixture by as much as 30 percent. The exact amount saved, of course, depends on the number

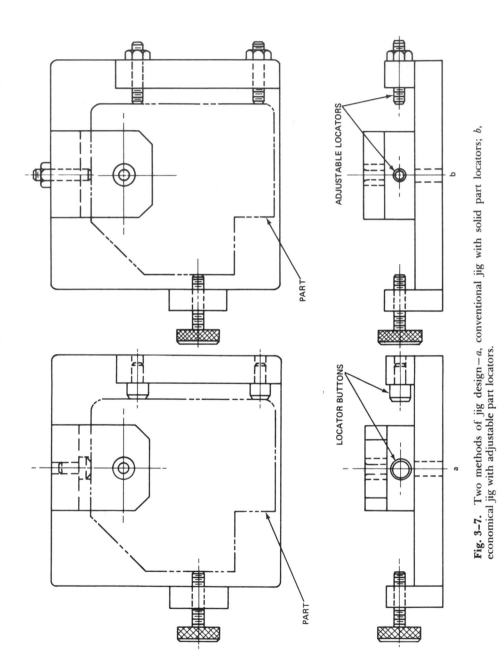

Fig. 3-7. Two methods of jig design — *a*, conventional jig with solid part locators; *b*, economical jig with adjustable part locators.

of locators, bushings, and other parts. Also, tools with adjustable locators can be easily adjusted to product changes, providing further savings. Finally, adjustable jigs and fixtures are not difficult to build and can be fabricated by unskilled personnel, relieving skilled toolmakers for more complex jobs.

Use of Basic and Standard Designs

Another method of reducing tooling costs is to use basic types of jigs and fixtures or low-priced standard components. Although these tools are simple and inexpensive, they will serve well for many low-volume production needs.

Template Drill Jigs. The simplest and least expensive drill jig is the template drill jig, which consists simply of a plate containing holes or bushings to guide a drill. This type of tool, not actually a true jig because it does not incorporate a clamping device, can be used for a wide variety of parts. The template drill jig is usually placed on the part for the purpose of drilling the desired hole pattern, and it may or may not have a means of locating the work. If it does not, it is oriented on the part by measurement.

Fig. 3–8 shows a simple template drill jig designed to locate two holes in a support bracket. The location of the holes is in relation to one edge of the bracket and a hole previously reamed into it. Because of its simple design and construction, this template jig was fabricated at a fraction of the cost that might have been expected. Savings were due to the design and to the use of pre-finished materials. Drill bushings were omitted because of the few parts to be drilled, and costs were further reduced.

Four more template jigs are illustrated in Fig. 3–9. A one-piece jig designed to drill two spaced holes in the end of a large shaft is shown in *a*. The jig was machined to fit over the shaft, and after the first hole was drilled a pin of proper diameter was placed in the hole to locate the jig for the orientation of the second

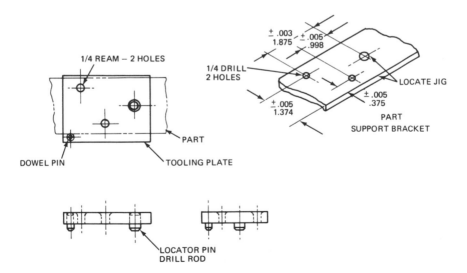

Fig. 3–8. Template jig constructed from prefinished material.

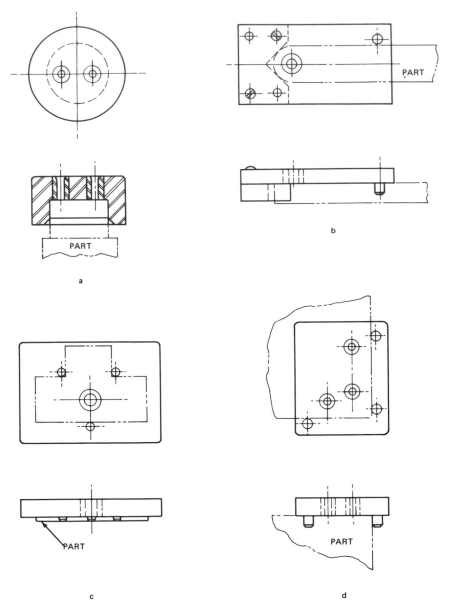

Fig. 3–9. Template jigs.

hole. Because of the weight and size of the shaft, a portable electric drill was used to drill the holes.

The jig shown in Fig. 3–9b consists mainly of a drill guide plate and a locating V block. It was used for drilling a hole into a lever, one of its requirements being that it produce the hole approximately in the center of the lever's rounded end.

The template jig shown in Fig. 3–9c was used for drilling a hole in a small sheet metal plate, the design of which required that the hole be located from the periphery of the plate. This was accomplished by means of three dowel pins press fitted into the jig as shown.

The jig illustrated in d is similar to that of c. In this case the jig was designed to drill three stud holes into the corner of a large machine base. A pin was placed into the first hole drilled to prevent the jig from moving while the remaining holes were drilled. Inserting a pin into the first drilled hole is standard practice when template drill jigs producing two or more holes are used.

Template drill jigs are generally used for large parts when holes are to be drilled in one portion of the part and a conventional jig to hold the entire part would be economically impractical. Template jigs are usually less expensive than conventional jigs, and it is often more economical to use two or three template drill jigs instead of one large or complex conventional jig.

Note that while the relationship of a hole pattern made with a template drill jig may not be accurately oriented to a part, the accuracy of the hole pattern within the template jig itself is equivalent to that of any conventional jig.

Plate Jigs. Plate jigs also consist of plates equipped with drill guide holes or bushings. Unlike template jigs, however, plate jigs incorporate a clamping system.

The plate jig's open construction facilitates part loading and unloading, locating, clamping, and chip removal. Most plate jigs, like template drill jigs, are inexpensive to design and fabricate.

Three different types of plate jigs are illustrated in Fig. 3–10. The plate jig shown in a is simply a template jig equipped with a clamping device. Its design is suitable for large parts.

Fig. 3–10b shows a plate drill jig made up of a plate with bushings, a locating stud, and standard screws used as jig feet. This type of jig is used for rela-

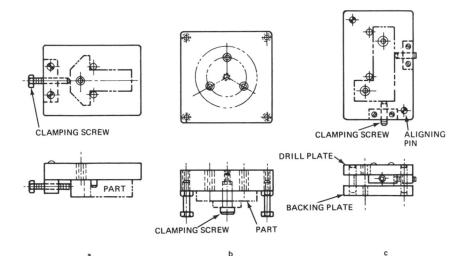

Fig. 3–10. Plate jigs.

tively small parts. Its size is restricted because it must be held by hand on the drill press table. It cannot be clamped to the table because it has to be inverted for loading and unloading.

Another plate jig, often referred to as the "sandwich" jig, is shown in Fig. 3–10c. This jig positions the part between a backup plate and a drill plate which contains the drill guide holes or bushings, the locators, and the clamps. The backup plate has clearance holes for the drill and is aligned with the drill plate by two pins. Use of this type of plate jig is limited to flat plate or sheet metal parts.

All clamping and locating devices for plate jigs should be based on the drill plate. This rule is mandatory for sandwich-type plate jigs in order to avoid unnecessarily accurate orientation requirements for the backup plate.

Commercial Universal Jigs. Universal jigs are manufactured as standard basic units and are sold in many different sizes and styles similar to that shown in Fig. 3–11. The main features of these jigs are rigidity, low height, ample chip

Fig. 3–11. Standard universal jig. (*Courtesy, Universal Vise and Tool Company*)

clearance, and ease of operation. They are essentially fast-clamping vises designed so that they can be easily altered to serve as drill jigs. Generally, all that has to be done to prepare a universal jig for use is to provide it with a work support or adapter. A typical universal jig prepared for drilling a hole into a small rod is shown in Fig. 3–12. This jig features an adjustable stop and a self-locating bushing liner. To decrease its height it was necessary to attach a plate to the base.

With the use of removable inserts, bushings, locators, etc., the universal jig can be adapted to a variety of workpieces. It is also reusable for other jobs although it may be necessary to replace the top plate.

Universal jigs are ideal for limited-production manufacturing because of their versatility and economy. One manufacturer claims that they can reduce

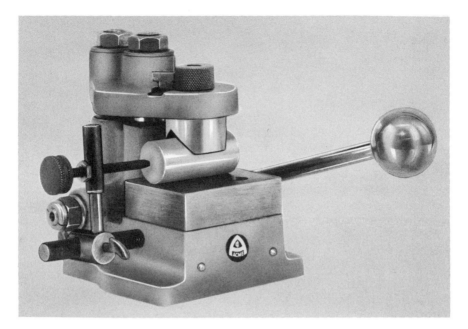

Fig. 3–12. Universal jig ready for production. (*Courtesy, Acme Industrial Company*)

tooling costs by one-third. In addition, they can be designed and adapted in a fraction of the time normally required to design and fabricate conventional jigs to perform the same functions.

Universal Rotary Table. Another useful device which is available as a standard unit and which is ideal for low-volume manufacturing is the universal rotary table, currently available with adapter plate, face plate, or three-jaw chuck as shown in Fig. 3–13. This particular table can be mounted to a machine

Fig. 3–13. Standard universal rotary table with (l. to r.) three-jaw chuck, face plate, and adapter plate. (*Courtesy, Universal Vise and Tool Company*)

bed in either a vertical or a horizontal position. The rotary table shown in Fig. 3–14 was mounted horizontally for a limited-production application on a milling machine. A rotary table can also be used as an indexing drill jig simply by mounting a bushing plate to the vertical base.

Machine Vises. Before deciding on special jigs or fixtures for a particular operation, the tool engineer should consider the use of a standard machine vise. The vises illustrated in Figs. 3–15 through 3–17 are representative of the types most suitable for use as jigs and fixtures, and they can all be bought at reasonable cost. The vises shown in Figs. 3–15 and 3–16 have swivel bases and are perhaps the most common types produced. For compound-angle machining

Fig. 3–14. Limited-production milling application of a rotary table. (*Courtesy, Universal Vise and Tool Company*)

Fig. 3–15. Swivel vise. (*Courtesy, Brown and Sharpe Company*)

or drilling, the inexpensive three-way vise in Fig. 3–16 is a useful workholding device. Often a single-purpose, compound-angle fixture will cost more, so justifying such a vise, even for limited production, should not be difficult.

For holding large parts, the mill vise shown in Fig. 3–17 is most suitable. Because of its patented locking principle and tie-bar arrangement, the only limit to the separation of the jaws is the length of the machine table. However, the vise is adaptable only to machines with T-slot tables.

The vises discussed, when modified with special jaws, can be both versatile and economical as workholding fixtures. Also, their jaws can be easily changed to meet different job requirements. With various types of V-jaw arrangements, cylindrical parts can be oriented and held securely. To hold

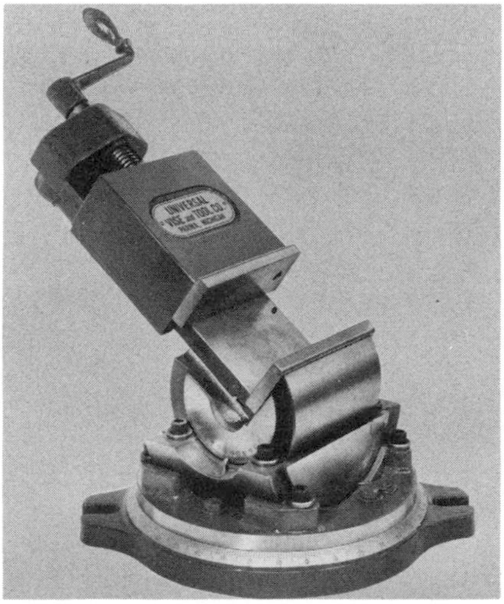

Fig. 3–16. Three-way vise. (*Courtesy, Universal Vise and Tool Company*)

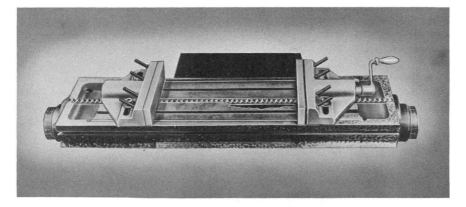

Fig. 3–17. Mill vise. (*Courtesy, Universal Vise and Tool Company*)

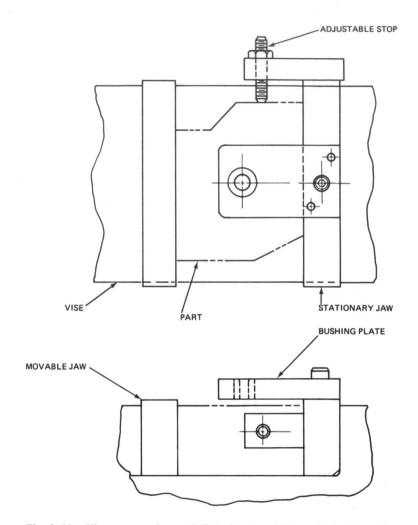

Fig. 3–18. Vise converted to a drill jig by the addition of a bushing plate and an adjustable stop.

castings and parts of irregular contour or to prevent marred finishes, leather or rubber can be bonded to the jaw faces. Extended jaws can also be adapted to any of the vises shown in Figs. 3–15 through 3–17 to increase their capacity in height or width. Valuable ideas on the design and application of special vise jaws are too numerous to mention here, but can be found in books on tool design listed in Additional Readings.

The usefulness of a commercial machine vise is virtually unlimited, for it can also be used as a drill jig merely by mounting a bushing plate to the stationary jaw as shown in Fig. 3–18. For additional versatility and economy when a variety of parts are to be drilled, several bushing sizes can be provided in a

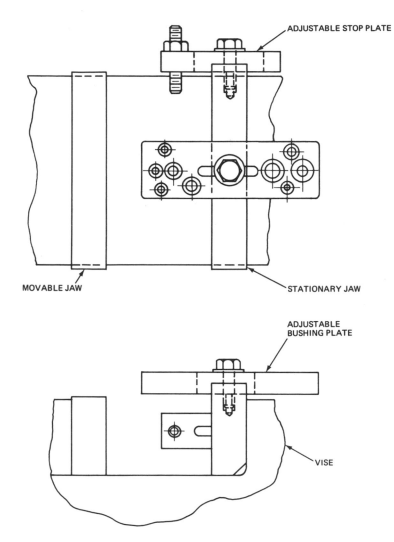

Fig. 3-19. Vise converted to a drill jig with an adjustable bushing plate.

single adjustable plate as illustrated by Fig. 3-19. This arrangement is especially suitable when production quantities do not warrant single-purpose drill jigs or when engineering changes are anticipated.

chapter
4

methods of
low-cost design

The principles of low-cost tool design discussed in the last chapter emphasize the methods of minimizing material costs and fabrication labor costs for jigs and fixtures. There is, however, yet another method of reducing low-production tool costs — through the reduction of many of the labor costs involved in the actual design of the tool.

Design represents approximately 30 percent of the total cost of a tool, and it is obvious that any reductions in design cost will lower total cost. This chapter examines the means of designing jigs and fixtures efficiently and eliminating a substantial percentage of their design costs.

DESIGNING A LOW-COST TOOL

The process of designing a tool — that is, putting the principles of low-cost design onto paper for the toolmaker to follow in fabrication — consists of two basic steps:

1) *The concept.* — The idea or concept of a jig or fixture to perform a job must, of course, be formed in the designer's mind.
2) *The drawing.* — After the designer has developed an idea for a needed tool, he must be able to transfer his idea to paper in the form of a working drawing. On paper, the idea becomes a permanent document that serves as a record of communication between designer and toolmaker. Obviously, the document should be legible and should conform to standard drafting practices in order to eliminate guesswork and misunderstanding on the part of the toolmaker.

Both these steps require time, so both must be minimized to reduce design costs. Shortening the first step, the creative period, can be very difficult. The ability to create or originate thoughts and develop them into a tool design is an ability which some designers have to a greater degree than others. There is no doubt that the ability to create can improve with special training and experience, and in many cases management has found that training can provide valuable savings. In general, however, any elimination of costs during the creative period is determined by the individual abilities of the designer. Therefore, the second step in the design process — transferring the tool concept to paper — is the most logical area in which to find tangible and immediate savings.

41

TOOL DRAWING METHODS

Drafting practices are a formalized routine. For that reason they are often overlooked as a source of savings. But many companies, with a small amount of retraining for their designers, have reduced drafting time and cost through new, time-saving techniques.

In design, tool drawings do not need to be formalized as do product drawings. Shading, fancy lettering, and unnecessary detailing may be eliminated without harming the purpose of the drawing. Often even details and views may be left out of the drawings with no restriction in communication of the tool idea from designer to toolmaker.

To better illustrate this point, a conventional drawing of a simple plate with screws and dowel pins, frequently used in jig and fixture construction, is shown in Fig. 4–1a. Fig. 4–1b shows the same plate drawn by a simplified, time-saving method. Note that it is only necessary to show the center lines to indicate the screw and dowel positions. The letters "D" and "S," placed at the center line locations, distinguish dowels from screws.

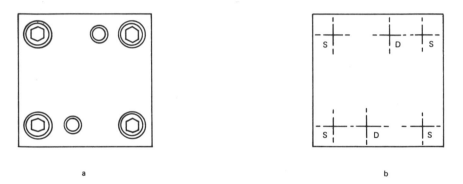

a   b

Fig. 4–1. Methods of drawing screw and dowel locations—*a*, conventional; *b*, simplified.

When many holes of various sizes are to be drilled, the method of drawing shown in Fig. 4–2 is satisfactory. Again, center lines are used for the hole locations, and letters designate the different hole sizes and types. Using the center line and letter method of indicating hole locations can result in considerable savings in time, particularly when many holes are involved.

A toolmaker is a highly skilled craftsman and can, and often does, work from drawings containing a minimum of information. It is up to the designer to decide what information is essential for the toolmaker to know and what is not. In many cases only an assembly view containing the necessary dimensions is sufficient. Precise location of screws, dowels, blocks, clamps, etc., may be unimportant, and if so, those parts can probably be scaled from the drawing or located by the toolmaker at his discretion. To save further time, the designer may omit from the drawing all standard commercial parts such as clamps, locators, and bushings. Fig. 4–3 compares the two methods of drawing tools

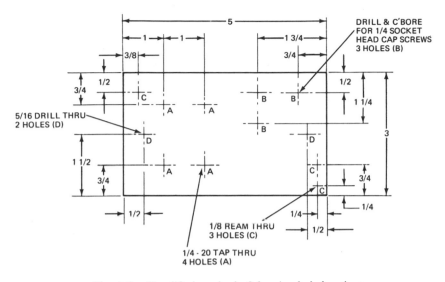

Fig. 4–2. Simplified method of drawing hole locations.

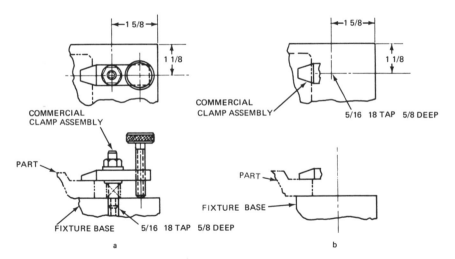

Fig. 4–3. Methods of drawing commercial parts — *a*, conventional; *b*, simplified.

which use standard components. That a large amount of drafting that can be eliminated by the simplified method is obvious.

The point is further demonstrated by the drawings shown in Figs. 4–4 and 4–5. A milling fixture was designed and drawn in both conventional and time-saving drafting techniques. The same designer drew both. The conventional method was favored in this experiment because it was drawn last, after the designer had become more familiar with the fixture design.

STOCK LIST

DET.	REQ'D	MAT.	FINISH STOCK SIZE	REMARKS
1	1	C. R. S	1/2 x 4 x 6 1/8	
2	1	T. S.	3/16 x 3/4 x 1 5/8	HDN. RC. 58 - 60
3	1	T. S.	7/8 x 1 5/8 x 3	HDN. RC. 58 - 60
4	2	DOWEL - 1/8 DIA. x 5/8 LG.		STD.
5	4	SOC. HD. CAP SCR. #8 - 32 x 3/4 LG.		STD.
6	1	DOWEL - 3/16 DIA. x 5/8 LG.		STD.
7	1	VLIER #NM - 52 N SPRING PLUNGER		STD.
8	1	#10 - 32 JAM NUT		STD.
9	2	SOC. HD. CAP SCR. #8 - 32 x 3/8 LG.		STD.
10	2	DOWEL - 1/8 DIA. x 3/8 LG.		STD.
11	2	STD. Parts# SLFK - 500 KEY		STD.
12	1	CARR LANE CLAMP ASS'Y # CL - 3520 - 2 - 1		STD.

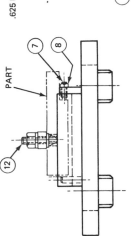
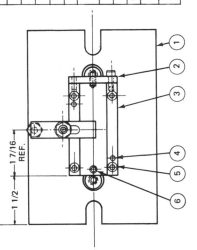
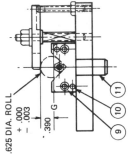

Fig. 4-4. Milling fixture drawn with conventional drafting methods. Other details of this drawing have been deleted because of limited space.

STOCK LIST

DET.	REQ'D	MAT.	FINISH STOCK SIZE	REMARKS
1	1	C. R. S.	1/2 x 4 x 5	
2	1	T. S.	3/16 x 11/16 x 1 1/2	HDN. RC. 58 - 60
3	1	T. S.	3/4 x 1 1/2 x 2 7/8	HDN. RC. 58 - 60
4	2		DOWEL - 1/8 DIA. x 5/8 LG.	STD.
5	4		SOC. HD. CAP SCR. #8 - 32 x 3/4 LG.	STD.
6	1		DOWEL - 3/16 DIA. x 5/8 LG.	STD.
7	1		VLIER # NM - 52 N SPRING PLUNGER	STD.
8	1		#10-32 JAM NUT	STD.
9	2		SOC. HD. CAP SCR. #8 - 32 x 3/8 LG.	STD.
10	2		DOWEL - 1/8 DIA. x 3/8 LG.	STD.
11	2		STD. PARTS #SLFK - 500 KEY	STD.
12	1		CARR LANE CLAMP ASS'Y #CL - 3520 - 2 -1	STD.

NOTES:
1. SCALE ALL DIMENSIONS NOT SHOWN
2. FOR INFORMATION NOT SHOWN WORK TO TOOLMAKERS STANDARD PRACTICE
3. UNLESS OTHERWISE SPECIFIED, LOCATE ALL SCREWS AND DOWELS TO SUIT.

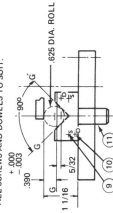

Fig. 4–5. Milling fixture drawn with simplified, economical drafting methods.

Drafting time for the conventional drawing shown in Fig. 4–4 was three hours and fifty minutes. Drawing the same fixture by the time-saving method illustrated in Fig. 4–5 required only two hours and forty-five minutes — a time reduction of about 28 percent. If the fixture had comprised more parts, the time savings would have been even more significant.

Three shortcuts that were used in Fig. 4–5 and that led to this savings in time were:

1) Elimination of the drawing details by inserting the dimensions directly on the assembly views

2) Elimination of many dimensions on the drawing by providing finished detail sizes in the stock list

3) Provision of notes which not only saved drafting time but allowed the toolmaker more latitude in fabrication.

Further savings were possible when the fixture was inspected, for the inspector needed only to check those few critical dimensions shown in Fig. 4–5. If the inspector had been required to work from the drawing in Fig. 4–4, all the dimensions would have to be inspected.

chapter

5

multipurpose jigs & fixtures

In spite of the tool engineer's best efforts to reduce the costs of individual jigs and fixtures for his limited production operations, he may find that their total cost still exceeds the desirable level and cuts into profits that might otherwise be made. When this problem occurs, the tool engineer may often be able to combine the features of two or more of his tools into a single jig or fixture. Such a tool, designed to be used for more than one operation or part configuration, is called a *multipurpose jig or fixture*. This chapter examines the applications of multipurpose jigs and fixtures, points out their limitations, and provides a method of estimating their costs and the savings they may provide.

MULTIPURPOSE TOOL APPLICATIONS

Before the tool engineer decides to tool up for limited production, he should first analyze the operations involved and the parts to be produced to determine if a multipurpose tool would actually be practical and economical. There are two major limitations of multipurpose tools that he must seriously consider before deciding to apply such tools to his own particular tooling problem. Multipurpose jigs and fixtures can be used:

    1) When parts produced or operations performed are similar
    2) When schedules and production quantities permit.

Similarities in Parts and Operations

The various parts to be made, and the operations involved in making them, should be carefully studied for any points of similarity. Similar parts or operations should then be segregated and examined to determine the features of their designs that can be worked with a single, cheaper jig or fixture. Radically different parts or operations are seldom suitable for multipurpose tooling; multipurpose tools that are complex enough to accommodate unlike parts or operations are usually impractical because of their high cost.

Schedules and Production Quantities

If the tool engineer finds similar operations that multipurpose tools can perform, he should make a thorough study of production requirements to see if the multipurpose tooling will be compatible with time schedules and quantity

47

requirements. Multipurpose jigs and fixtures should be used when production commitments can be handled through part-time or interrupted operations, but not when operations must be run continuously or simultaneously.

For example, assume that Part A and Part B are similar and a multipurpose tool is being considered for their production. Eighty units of each part are needed. Part A requires 1 hr/unit and Part B .8 hr/unit for production. Part A can then be produced in 80 hours and Part B in 64 hours. If only 80 hours' total production time can be allowed for both parts, however, it is obvious that the operation for Part A must be run continuously and that for Part B simultaneously in order to meet the production deadline. A multipurpose tool probably could not be used to produce the two parts at the same time unless it was overly complex and costly.

If, on the other hand, the total production time for the two parts is not limited, then a multipurpose tool could be used to produce all units of A followed by all units of B or, if necessary, lots alternating A-B-A-B-A-B, etc. Total production time for the two parts would then equal 144 hours plus the amount of time required to convert the tool from the Part A operation to the Part B operation—and back if necessary.

Each process of converting a multipurpose jig or fixture from one use to another is called *changeover*, and the amount of time that process requires is called *changeover time*. Changeover, as we shall see next, has an important bearing on estimating the applicability and cost of a multipurpose tool.

MULTIPURPOSE TOOL COST ESTIMATING

Although multipurpose tooling may at first appear to offer cost advantages, the costs of setups or changeovers for different parts may cancel out many of those gains. Maximum savings are only possible when a part can be completed in a single setup, and each additional changeover subtracts from those savings. So before capital is invested on multipurpose tools, a simple cost study should be made.

A multipurpose tool cost estimate requires that the following data be known:
1) The cost of the multipurpose tool, including its design costs
2) The total cost of all the tools that the multipurpose tool is to replace, including their design costs
3) The time required for changeovers of the multipurpose tool
4) The number of changeovers, as determined from production schedules.
The formulas to be used in the cost estimate are as follows:

$$E = C \times D \tag{1}$$

$$F = B - (A + E) \tag{2}$$

$$G = \frac{B - A}{D} \tag{3}$$

Where:
  A = Cost of multipurpose tool, including design
  B = Total cost of replaced single-purpose tools, including design
  C = Number of changeovers necessary to complete the order

D = Cost of each changeover, including labor and overhead
E = Total cost of changeovers to complete the order
F = Savings through use of the multipurpose tool
G = Maximum number of changeovers up to the break-even point. If the number of changeovers is greater than G, then single-purpose tools should be used.

Example: For a certain machining operation on two similar parts, two separate drill jigs can be designed and fabricated for a cost of $475 for one and $325 for the other—a total of $800. A multipurpose drill jig for both parts can be designed and constructed for an estimated $620. Production schedules call for five lots to be produced of each part. Five lots of two parts each will require ten operations—nine tooling changeovers. Changeover time is estimated as .6 hours at a cost of $7.00 per hour including labor and overhead. Each changeover will therefore cost $7.00 × .6, or $4.20. With this information, determine whether the multipurpose drill jig would be more or less economical than the two single-purpose jigs.

$$E = C \times D \tag{1}$$

$$= 9 \times \$4.20$$

$$= \$37.80 \qquad \text{Total cost of the changeovers}$$

$$F = B - (A + E) \tag{2}$$

$$= \$800 - (\$620 + \$37.80)$$

$$= \$800 - \$657.80$$

$$= \$142.20 \qquad \text{Possible savings through use of the multipurpose drill jig.}$$

$$G = \frac{B - A}{D} \tag{3}$$

$$= \frac{\$800 - \$620}{\$4.20}$$

$$= \frac{\$180}{\$4.20}$$

$$= 42.9 \qquad \text{Maximum number of changeovers to the break-even point.}$$

The results of this cost analysis indicate a savings of $142.20 in favor of the multipurpose drill jig. If more than 42 changeovers were necessary, however, the advantage of lower cost would be with the two separate, single-purpose jigs.

It is important to know the changeover break-even point before a decision on the type of tools to be used is made, for production schedules can often be adjusted to lower tooling costs by reducing the number of changeovers. Also, the break-even point may determine whether multi- or single-purpose tools should be used if future reorders are possible.

DESIGNING MULTIPURPOSE TOOLS

Two applications of multipurpose tools are described below. These examples show the principles of design and the capabilities of this type of tooling when either similarity of parts or similarity of operations makes their use feasible.

Combining Operations

The motor bracket shown in Fig. 5–1 was considered an ideal part for multi-purpose-type tooling because all the operations in its fabrication could be per-

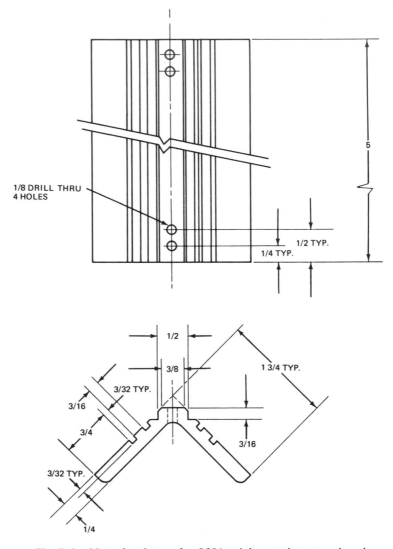

Fig. 5–1.   Motor bracket made of 304 stainless steel structural angle.

formed by one fixture without seriously affecting schedules. Fig. 5–2 shows the manufacturing process sheet calling for a multipurpose fixture for construction of the mount.

The result, a combination milling fixture and drill jig shown in Fig. 5–3, consists of a base (A) with holes reamed to accommodate two patented, removable reamed-hole keys (B). The holes are reamed in two planes to give the keys two mounting positions. The fixture is attached to the milling machine table by means of a socket cap screw and T-nut assembly (C) for which a second hole in another plane is also provided. A dowel (D) is press-fitted into the body of the fixture to orient the part. Four standard hook clamps (E) are used to secure the part. A detachable bushing plate (F) is provided to guide the drilling of the $1/8$-in holes.

MANUFACTURING PROCESS SHEET

Part Motor Bracket	Part No. 4861	Written by C. Daigle	Date 6/28/68	Rev. No. 0
Oper. No.	Operation	Machine	Tool	Remarks
010	Cut stock to 5 inch lgth.	Abrasive saw		
020	Deburr		File	
030	Mount part in fixture, mill 3/8 flat and 1/2 x 3/16 step	Milling machine	Fixt. 3040 Cutters 2035	Remove bushing plate from fixture 3040
040	Deburr		File	
050	Mount part in fixture mill (2) 3/32 slots	Milling machine	Fixt. 3040 Cutters 2036	Relocate keys and lock screw
060	Reverse part in fixture and repeat milling (2) 3/32 slots on opposite side	Milling machine	Fixt. 3040 Cutters 2036	
070	Mount part in fixture and drill (2) 1/8 holes	Drill press	Fixt. 3040 1/8 Drill	Remove keys, T-Nut and lock scr. Mount bushing plate
080	Reverse part in fixture and drill (2) 1/8 holes in opposite end	Drill press	Fixt. 3040 1/8 Drill	
090	Deburr all edges		File	

Fig. 5–2. Manufacturing process sheet for the motor bracket shown in Fig. 5–1.

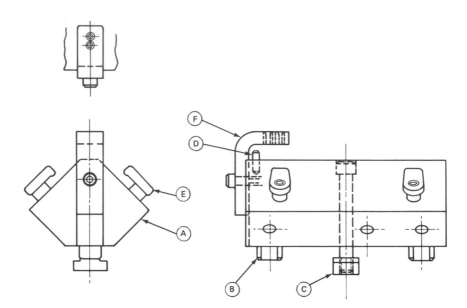

Fig. 5–3. Multipurpose fixture designed for milling and drilling the motor bracket shown in Fig. 5–1.

The mount is first cut to finish length and deburred. Then, for the flat milling operation (operation 030 on the process sheet), the bushing plate is removed from the fixture. With the part mounted in the fixture, and the fixture on a milling machine, the flats are produced by a set of gang cutters as shown in Fig. 5–4. For milling the slots (operations 050 and 060), the patented keys and the locking screw assembly are removed and relocated in their alternate holes as illustrated in Fig. 5–5. After gang cutters mill the slots on one side, the part is reversed on the fixture to machine the slots on the opposite side.

To drill the four .125-in diameter holes (operations 070 and 080), the fixture must be converted into a drill jig. This is done by attaching the bushing plate and removing the clamp screw assembly and reamed-hole keys as shown in Fig. 5–6. The first two holes are then drilled, and after the part is reversed the two holes on the opposite end are drilled.

Combining Similar Parts

As discussed earlier, similar parts and operations should be segregated and analyzed to determine the possibility of using one fixture instead of several. Excellent examples of similar parts for which a multipurpose tool was designed are the three links illustrated in Fig. 5–7, which are quite similar but which have minor differences in dimensions.

The links could be cut to length and formed without difficulty by using standard tooling and equipment, and the holes could be drilled with a conventional, single-purpose drill jig for each link. Note, however, that the holes in each link are the same diameter. This fact indicates that it is possible to use one

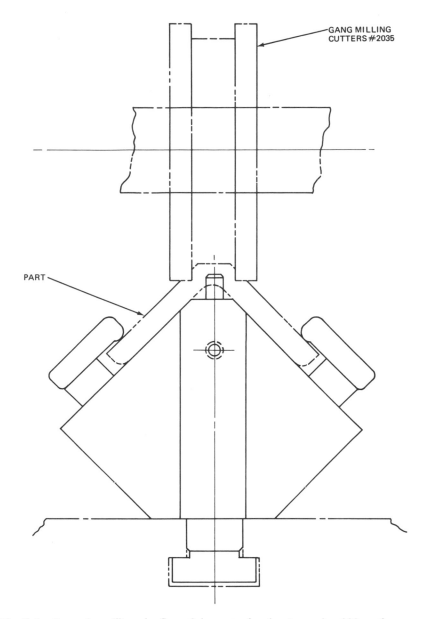

GANG MILLING
CUTTERS #2035

PART

Fig. 5-4. Setup for milling the flats of the motor bracket (operation 030 on the process sheet).

multipurpose drill jig for all three links. The multipurpose jig that was designed for the links is shown in Fig. 5–8.

The jig incorporates a removable bushing plate (A) which can be altered to accommodate all three links by changing its position with the screws and

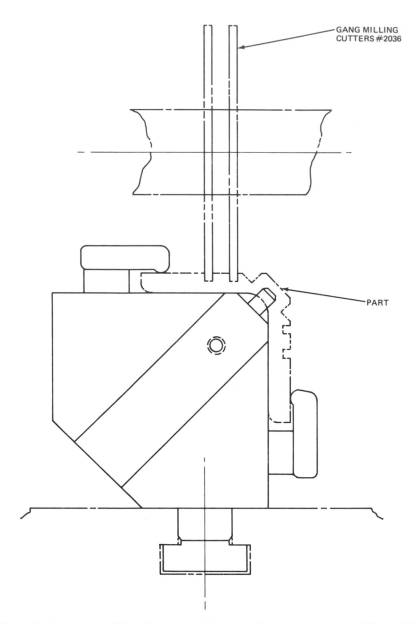

Fig. 5–5. Setup for milling the slots of the motor bracket (operations 050 and 060).

dowels provided in the jig base. The .5-inch and .25-inch spacers (B) are fastened to the bushing plate to allow its adjustment for the 1-inch- and 1.5-inch-wide links. A spacer is not required for the 2-inch-wide link. The #29 (.136-in diameter) holes, being located alike in the ends of all the links, do not

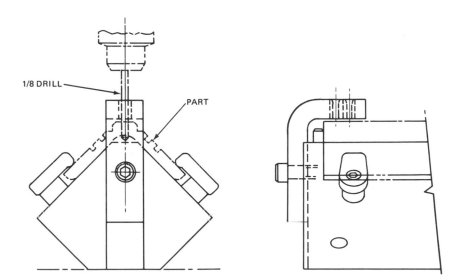

Fig. 5-6. Multipurpose fixture converted into a drill jig for operations 070 and 080.

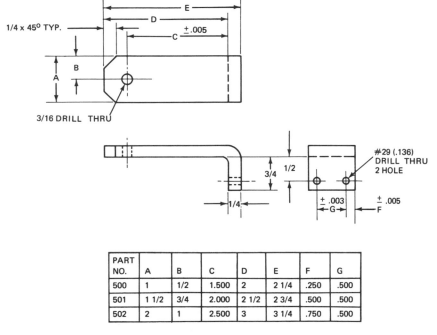

PART NO.	A	B	C	D	E	F	G
500	1	1/2	1.500	2	2 1/4	.250	.500
501	1 1/2	3/4	2.000	2 1/2	2 3/4	.500	.500
502	2	1	2.500	3	3 1/4	.750	.500

LINK—MATERIAL 1020 STEEL

Fig. 5-7. Drawing for three steel links.

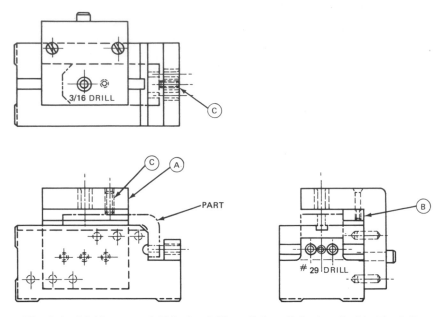

Fig. 5–8. Multipurpose drill jig for drilling all three links described in Fig. 5–7.

require an adjustable bushing plate. For drilling these two small holes, the drill jig is simply rotated on end. Two soft-pointed socket set screws (C) are used to clamp the links.

ADDITIONAL ADVANTAGES OF MULTIPURPOSE TOOLS

Multipurpose tooling offers a number of advantages in addition to lower cost. The production planner should examine these advantages closely for suggestions on ways he might make production easier and more economical. These advantages may provide opportunities for improvements in any of the following areas:

1) *Accuracy.*—Multipurpose jigs and fixtures provide accuracy as well as low cost. Their accuracy is generally equal to equivalent single-purpose tooling, and when more than one operation is performed on the same part they may even be more accurate, since no errors are introduced because of slight differences between tools. However, tolerance specifications of multipurpose jigs and fixtures, like those of single-purpose ones as explained in Chapter 3, should be no closer than 40 percent of part tolerances for the best economy.

2) *Space requirements.*—Substitution of one tool for many offers an additional advantage to shops or factories where storage or operation space is limited. Although space may be of little importance when tools of moderate size are used, large fixtures can take up valuable working or storage space and become a major problem. In extreme cases, even complex

multipurpose tools may be desirable for their smaller space requirements in spite of their economic disadvantages.

3) *Lead time.* — Multipurpose tooling should also be seriously considered when a product must be manufactured with minimum lead time. Generally, multipurpose tooling can be designed and fabricated in a fraction of the time required for a conventional program involving several separate tools.

4) *Component interchange.* — In many cases, if production schedules warrant, some of the cost-saving advantages of multipurpose tools may be gained by the interchange of standard tooling components such as clamps, bushings, pins, etc., between single-purpose tools. This system may offer the tool engineer more flexibility in planning production lines for which full multipurpose tooling is not feasible.

5) *Mass production.* — Although jigs and fixtures of multipurpose design have proved most successful and economical for limited-production manufacturing, even large production quantities can often be handled easier and cheaper with tools that perform more than one operation. When the number of changeovers and total changeover time can be held at low levels for extended runs, the total cost of changeovers can be lowered and the number of changeovers to the break-even point can be increased.

chapter
6

universal tooling systems

Universal or *erector-set* tooling systems are kits of standard tooling components or subassemblies which may be bolted together in a variety of combinations as jigs and fixtures. When short lead time or small production quantities do not warrant the fabrication of special tooling, a universal tooling system (called "erector-set" tooling because of its resemblance to the popular children's construction toy) may make production profitable. The costs of these kits may at first seem prohibitive but are a result of the design and engineering that make them adaptable and accurate. Because of the adaptability, accuracy, and reusability of the systems, their initial costs can generally be absorbed during their first year of use.

In this chapter we discuss the design, construction, recording, and advantages of jigs and fixtures assembled from universal tooling systems representative of the many types of kits that are commercially available.

DESIGNING ERECTOR-SET TOOLING

Tool design is virtually eliminated when a universal tooling system is used. Formal design drawings are not required because erector-set jigs and fixtures can be built from the directions of operation data sheets and part drawings prepared by process engineers. The process engineer simply indicates the areas of the workpiece to be machined and those to be used as locating surfaces; the method of supporting and clamping the workpiece is the responsibility of the tool assembler.

A typical process operation drawing with all necessary information is shown in Fig. 6–1. Using this drawing, the tool assembler can quickly and easily construct a jig from the elements of the universal tooling kit. If operation drawings are unavailable, or if lead time must be reduced to meet production schedules, freehand sketches of the part can be used successfully.

Since erector-set tooling does not require design drawings, engineering changes to the tool can be processed quicker and more efficiently – when the tool assembler is notified of the change, the tool is immediately modified to incorporate the revision. This rapid response would not be possible with a conventional tooling program in which the program tool engineer must be notified of the change, must revise all the tool design drawings that are affected, and must then forward revised tool drawings to the toolroom where the neces-

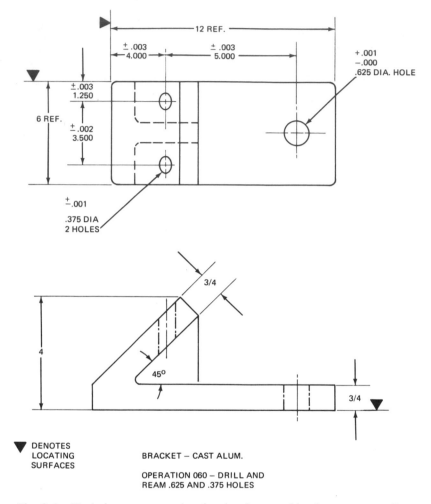

Fig. 6-1. Typical process operation drawing for assembly of an erector-set jig.

sary change to the tool is made. A conventional tool could be further delayed if modifications demand procurement of raw material or commercial components, a situation which would be unlikely to occur with an erector-set jig or fixture.

CONSTRUCTING ERECTOR-SET TOOLING

Universal tooling kits are composed of base plates, spacers, locators, bushings, stop elements, clamps, T bolts, and many other parts as shown in Figs. 6-2 and 6-3. These interchangeable components are so accurately machined that a jig or fixture maintaining tolerances to .0003 in. can be constructed from them.

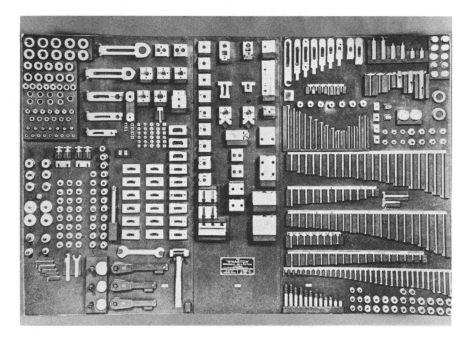

Fig. 6–2. Typical universal tooling system for both jigs and fixtures. (*Courtesy, Wharton Unitools*)

The first step in erecting a jig or fixture from the universal tooling kit is to select a base of suitable size. Standard bases provided with the kits are machined and ground to a high degree of accuracy, usually from solid nickel-steel castings. T slots accurate in width and pitch to .0003 in. are machined in the bases for the attachment of other tooling components.

After the base is selected, the general configuration of the tool is built up with the other tool elements, which are supplied in many shapes and sizes. Blocks or bars are used to construct the main part of most tool superstructures, and stop-thrust elements are added when necessary for reinforcement. Finally, more specialized elements are added to adapt the tool to its particular configuration and function. A typical fixture assembled from an erector-set kit is shown in Fig. 6–4*a*, and Fig. 6–4*b* shows a part mounted in it.

Construction with Sample Parts

The ideal method of construction with a universal tooling kit is to build the tool around a sample part. The sample part is placed on the jig or fixture base, and clamps, locators, etc., are added as they are needed. By this procedure, construction time can be reduced up to 50 percent. Often, if the sample part is accurately machined, the jig or fixture can be erected without measuring instruments. In some cases it is economical to fabricate a single part for the sole purpose of facilitating complex tool construction. Use of a sample setup part can also make frequent assembly and disassembly of jigs and fixtures economical.

Fig. 6–3. Typical erector-set tooling kit for fixtures. (*Courtesy, Kearney & Trecker Corporation*)

Construction with Templates

If a sample part is not available, an erector-set jig or fixture may be built around a template of the part. Examples of this technique are shown in Fig. 6–5. Both the drill jig shown in Fig. 6–5a and the milling fixture in b were constructed with the aid of the same part template. Templates can also, to some degree, be useful in determining if a part can be loaded or unloaded without interference. Templates undoubtedly simplify tooling construction and should be used whenever assembly problems arise.

Machining

Machining is seldom necessary when jigs and fixtures are assembled from universal tooling systems, although limited machining may occasionally be required to produce additional special components. Excessive machining or modification should always be avoided, however, for this will defeat the economic advantages of the system. If it is impossible to construct an erector-set jig or fixture without considerable machining, the universal tooling system should not be used.

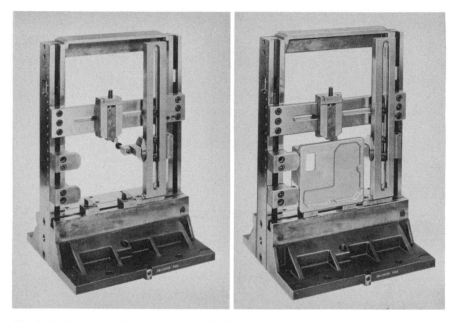

Fig. 6–4. Erector-set fixture—l., assembled for use; r., with a workpiece mounted. (*Courtesy, Kearney & Trecker Corporation*)

Fig. 6–5. Template used in the construction of erector-set tooling—l., drill jig; r., milling fixture. (*Courtesy, Wharton Unitools*)

Assembly Personnel

Fig. 6–6 shows an erector-set fixture being constructed. The ease of assembly of such tooling does not eliminate the need for knowledge and experience on the part of the assembler, for more than a mere tool shape is required.

Fig. 6–6. Fixture being assembled from an erector-set tooling kit. (*Courtesy, Welch Scientific Company*)

Jigs and fixtures constructed from these systems must be sufficiently strong to withstand the various machining forces that will be imposed on them during their use. In many cases, they must also be built to incorporate adjustable components, and they must often accommodate the configuration of an in-process part. The success or failure of an erector-set tooling program depends largely on the foresight, tooling ability, and imagination of tool builders who, with experience, can foresee and make reservations for contingencies; careful consideration should be given to their selection.

RECORDING ERECTOR-SET TOOLING

After an erector-set jig or fixture is constructed and before it is released to production, it should be photographed, and all its components should be listed. The list and the photographs should then be filed for future reference. With the photographs and information as guides, a jig or fixture can usually be reassembled in less than half its original construction time. The photograph file can also provide much basic information for assistance in developing permanent production tools.

ADVANTAGES OF ERECTOR-SET TOOLING

Because of the inherent advantages of universal tooling systems, no progressive company can afford to overlook them in its search for low-cost tooling. Companies engaged in limited or diversified production should seriously consider such systems because of their adaptability and reusability as well as the reduced design costs and short lead time they provide. Merely having "backup" tooling available for emergencies is well worth investment in an erector-set kit.

Reduced Lead Time

The major advantage of a universal tooling system is that it reduces tooling lead time and construction time. Reductions of up to 80 percent over conventional tooling times are not unusual. As an example of these savings, one erector-set milling fixture required a lead time of $2\frac{1}{2}$ days and a construction time of 7 hours. It was estimated that an equivalent conventional fixture would have required $3\frac{1}{2}$ weeks' lead time and 42 hours' fabrication time plus a material cost of $200, for a total cost of over $500.

One well-known instrument manufacturing company estimates that 9,000 tooling hours were saved in one year through the use of a universal tooling system. The company's average lead time for erector-set jigs and fixtures was 2½ days, and their average construction time was 7 hours.

Adaptability

Universal tooling kits can easily be adapted for production of entirely different products as well as similar products which require continual minor modifications. This adaptability especially recommends the systems to companies which experience frequent product changes. New products will not be held up because of tooling delays, and urgently needed modifications can be immediately incorporated into the tooling without interrupting normal production.

Experimental or prototype tooling can be constructed rapidly and inexpensively with universal tooling systems. When a complex permanent jig or fixture must be designed and built, it is often economical to first construct an erector-set model. It is much easier for a tool designer to derive the final concept of a permanent tool from such a working model.

Reusability

In spite of its high initial cost, the erector-set system can actually reduce the total tooling investment required over a period of one year or more. Once a production run is completed, an erector-set tool can be dismantled and the separate component parts can be returned to the tool crib for reuse, whereas a conventional tool becomes virtually worthless and must be stored or sold for scrap metal at a fraction of its original cost. Storage of conventional tooling for future use may often be considered, but storage, when coupled with additional expenses for occasional rehabilitation, is costly.

Backup Ability

The universal tooling system makes an excellent temporary replacement for permanent tooling which may have to be removed from production for repairs or modifications. It is comforting to the manufacturer to know that a backup tool is always available to replace any permanent production tool that might break down.

chapter

7

formed-section tool construction

In recent years, standard formed sectional material has become increasingly popular in the construction of low-cost jigs and fixtures. This is due to the savings in both time and cost that such material provides in the design as well as the fabrication stages of tool building.

Two types of standard formed material are used most frequently in tooling construction. They are:

1) Standard commercial premachined sections or forms (tooling plate)
2) Structural forms.

This chapter describes the uses of both these types of materials, points out the advantages of each, and provides examples of actual applications.

COMMERCIAL PREMACHINED SECTIONS

A wide variety of tooling plate forms and sizes are commercially available for tooling construction. Some of the standard premachined forms are shown in Fig. 7–1. Most of the shapes are available in 25-in lengths; the V section is available only in 16-in lengths. All the forms are made of normalized cast iron or aluminum which is machined square and parallel within .005 in/ft on all surfaces except the ends. Additional machining by the toolmaker is not normally required to qualify tooling plate surfaces, and several shapes can be fastened together with screws and dowels only.

A typical drill jig constructed from premachined sectional material is shown in Fig. 7–2. A standard commercial T section is used for the base, and a bushing

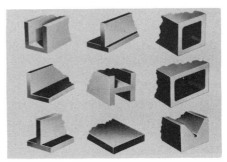

Fig. 7–1. Standard commercial sectional shapes used for tooling. (*Courtesy, Standard Parts Company*)

Fig. 7–2. Drill jig made from standard T-section tooling material. (*Courtesy, Standard Parts Company*)

plate made from flat tooling plate is fastened to it with screws and dowels. The jig is completed by adding standard jig feet, drill bushings, and clamps.

Design Simplification

Design drawings of tools made from premachined sectional material may consist simply of free-hand perspective sketches like the one of the milling fixture shown in Fig. 7–3. Formal drawings are seldom needed unless the tooling is complex or nonstandard components are used. In any case, design sketches should be drawn neatly and should contain all information that the toolmaker needs.

Reduction of Lead Time

Because the use of sectional material speeds the design and fabrication of tooling, it is particularly useful when products must be manufactured with minimum lead time. Not only is design simplified and speeded up, but fabrication time is reduced. For example, the drill jig shown in Fig. 7–4 was constructed from U-shaped premachined cast iron section and flat plate in only 9.5 hours, saving 30 percent in construction time.

Construction time for the milling fixture shown in Fig. 7–5 was 3.5 hours. The use of flat plate provided savings of 10 percent over conventional construction time. The grinding fixture illustrated in Fig. 7–6 was made of T-section tooling material and required only four hours' construction time — a savings of 40 percent.

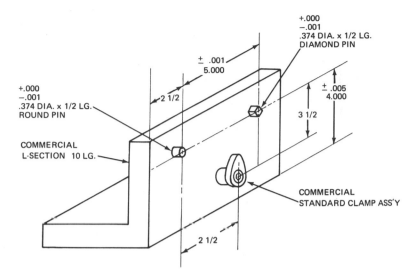

Fig. 7–3. Milling fixture made from standard L-section tooling material.

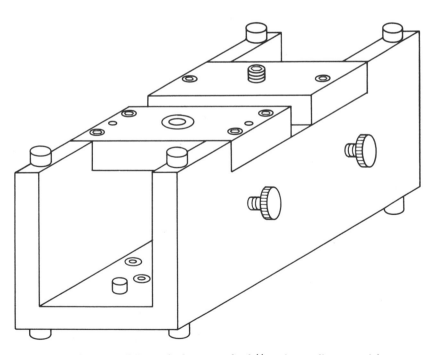

Fig. 7–4. Drill jig made from standard U-section tooling material.

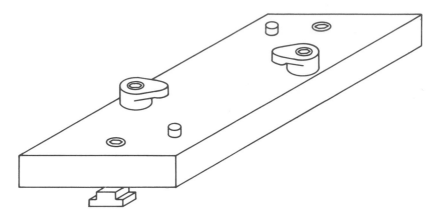

Fig. 7–5. Milling fixture made from flat tooling plate.

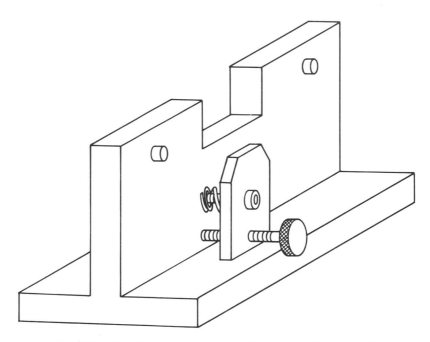

Fig. 7–6. Grinding fixture made from T-section tooling material.

Cost Savings

Premachined sectional forms can provide savings of as much as 50 percent of the cost of conventional tooling materials. When weldments and casting are specified for conventional tools, the savings provided if tooling plate is used instead may reach over 60 percent, and one manufacturer claims 75 percent savings on a particular application.

Savings on design costs can also be substantial when standard sectional tooling forms are used; conservatively, they can be estimated to range from 5 to 20 percent.

STRUCTURAL FORM MATERIAL

Structural form stock also offers useful benefits in tooling construction. It is available in many forms and various materials, and, in contrast to standard sectional tooling plate, in random lengths up to 60 ft. The greater lengths, of course, are ideal for large tooling.

For parts or operations that do not require great accuracy, structural material for tooling need not be premachined. However, for accuracy, structural material of "selected quality" must be chosen and must be machined to the requirements.

An excellent example of the use of structural material for tooling is illustrated in Fig. 7-7. The drill jig shown consists of two pieces of angle iron

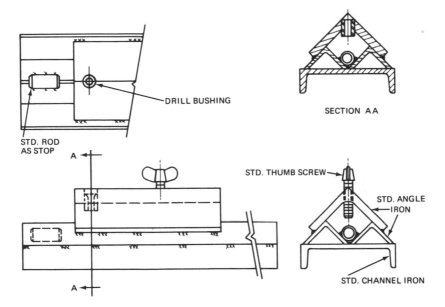

Fig. 7–7. Drill jig made from standard structural form material (channel and angle iron).

welded to a base of channel iron to form a V section. A larger angle iron, which serves as bushing plate and clamping mount, was welded over the V section. A thumb screw, a drill bushing, and a section of rod were added to complete the jig.

Since only limited accuracy was required, machining on the jig consisted solely of drilling and tapping for the thumb screw and boring a hole for the bushing. The jig was fabricated and ready for production in less than three hours.

To reduce tooling costs for the milling of the T slot and the drilling of four holes in the part shown in Fig. 7–8, the combination drill jig and milling fixture shown in Fig. 7–9 was devised. A structural H beam was used for the body

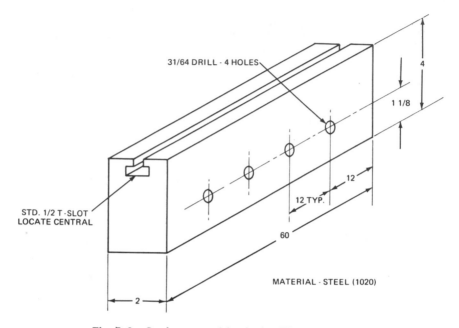

Fig. 7–8. Steel part requiring both milling and drilling.

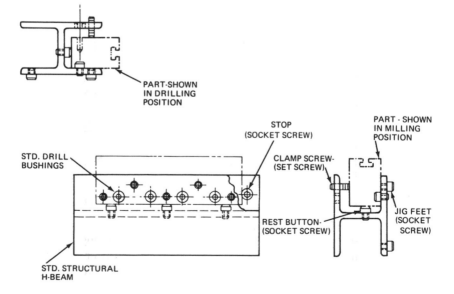

Fig. 7–9. Combination drill jig and milling fixture made from structural material (H beam) for the part shown in Fig. 7–8.

of this dual-purpose tool, and standard socket set screws were used as clamps. Cap screws served as jig feet, rest buttons, and a stop. Construction time for the combination tool was only 4.5 hours, certainly much less than would have been required for a comparable conventional tool.

Cost Savings

Structural material shows best economy when it is used for large tools. The potential savings of jigs and fixtures constructed from structural material are often proportional to tool size—the larger the tool, the larger the savings. In general, tooling costs can be reduced almost 25 percent through the use of these materials.

Savings are also increased by the simplification of design and the reduction of design costs for structural-material tools. Although the exact amount of such savings may be difficult to establish, it should not be overlooked.

chapter

8

magnesium
tool construction

Magnesium, one of nature's most plentiful metals, is light, strong, and easy to machine. These and other useful properties make it an excellent material for low-cost tooling. The first section of this chapter describes the properties of magnesium and three of its common alloys. The next section points out the chief disadvantage of the metal; if it is not handled carefully there is serious danger of fire. Special procedures are given to control magnesium fires if they do occur.

Finally, the last section describes how the commonly available forms of magnesium are used in making low-volume jigs and fixtures. Comparisons are made between the use of magnesium and other metals, particularly aluminum and steel plate. Special emphasis is given to tolerance and cost considerations.

CHARACTERISTICS OF MAGNESIUM

Magnesium, like many other metals, does not perform well as a structural material in its pure state, but must be alloyed for full effectiveness. When small percentages of aluminum, zinc, or other elements are present in a magnesium alloy, the physical properties of the metal are noticeably improved although weight is increased only slightly. Table VIII-1 shows some of the properties of three common magnesium alloys compared with the pure element. These and many other alloys of magnesium can be worked and shaped by practically

Table VIII-1. Properties of Magnesium and Three of Its Alloys.

	Compo- sition %	Tensile Strength (1,000 psi)	Elonga- tion % in 2 in.	Modulus of Elasticity	Brinell Hardness
Mg	99.9	14	5	6,500,000	30
AZ 31B	Al 3.0 Zn 1.0	42	15	6,500,000	73
AZ 61A	Al 6.5 Zn 1.0	45	16	—	60–70
ZK 60A	Zn 5.5 Zr .7	53	11	—	75–82

all known machining methods and have high strength-to-weight ratios, good thermal conductivity, excellent vibration damping properties, and other characteristics that make them ideal for low-cost tool fabrication.

Light Weight

Magnesium's most obvious and noted advantage is its light weight. One cubic foot of magnesium alloy weighs only about 112 lbs, compared to 165 lbs for the same volume of aluminum and nearly 500 lbs for steel. This relatively low weight of magnesium provides for labor cost reductions, since jigs and fixtures made from the light metal enable fewer workers to handle the same amount of work. At the same time, hoists and heavy lifting equipment are freed for other use.

Fig. 8–1 shows a magnesium alloy tumble jig that saved a leading aircraft manufacturer approximately 100 lbs in tool weight. It weighs 30 lbs rather than

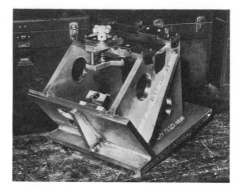

Fig. 8–1. Magnesium tumble jig. (*Courtesy, The Dow Chemical Company*)

the 130 lbs of a similar steel jig, and even at that, much of its weight results from the cast iron and steel bushings, pins, knobs, and fasteners used in its construction. The steel jig it replaced could be lifted only with a hoist, but the magnesium jig can easily be lifted manually by one man. Lightweight magnesium can also improve small tools that are frequently used and create a worker fatigue problem. Further savings can be realized simply because less expensive handling equipment is needed to handle magnesium-alloy tooling.

Machinability

Magnesium is about four times easier to work than most other tool construction metals. Its free-cutting qualities allow deeper cuts to be made at higher speeds and feeds and generally without liquid coolants. In a rough milling operation using high-speed-steel tools, for example, magnesium can be milled at a rate of 600–800 sfpm (surface feet per minute). The rates for other tooling materials are as follows:

Metal	Milling Rate (sfpm)
Aluminum	400
Steel	40–100
Cast iron	50–60

In addition, heavy-duty machine tools are not required for machining magnesium. Comparative power requirements for machining other metals, with magnesium serving as the unit base, are:

Metal	Power Requirement
Aluminum	1.8
Iron	3.5
Steel	6.3
Nickel	10.0

Because there is little or no tendency for the magnesium to tear or drag under machining operations, an excellent surface finish is possible; a surface of 3 to 5 microinches, for example, is easy to achieve. Close dimensional tolerances can be obtained in magnesium by conventional machining methods, eliminating many costly finishing cuts.

Strength

In applications for which strength and rigidity are requirements, magnesium offers another important benefit — high strength/weight and stiffness/weight ratios. Alloys of magnesium have specific tensile strengths (tensile strength ÷ specific gravity) approaching 30,000, whereas those of most stainless steels are less than 25,000. This means that it is possible to increase the thickness of magnesium tools slightly and thereby increase their strength to that of stainless steel while keeping their weight below that of similar stainless steel tools. As an example of this fact, consider a bar of magnesium and a bar of steel, each 1 sq in. in cross-sectional area and 10 in. long. The magnesium bar weighs .65 lb and has a tensile strength of 45,000 lbs. The steel bar weighs 2.83 lbs and has a tensile strength of about 100,000 lbs. If the cross-sectional area of the magnesium bar is increased to 3 sq in, its weight will be only 1.95 lbs, but its tensile strength will be increased to 135,000 lbs.

In much the same way we can compare the stiffness of magnesium to the stiffness of steel. Since stiffness of a structural part is directly proportional to the cube of its depth but weight is only proportional to the first power of the depth, an increase in the depth, or thickness, of a magnesium tool part will increase the rigidity and durability of the part without greatly compounding its weight.

Weldability

Magnesium can be easily welded by either inert-gas-shielded tungsten arc or metal arc welding. Oxyacetylene welds may be made on magnesium, but because of the difficulty of removing the flux required in the gas process, the arc welding methods are preferable. Either helium or argon, or a mixture of the two, can be used as the protective inert-gas atmosphere in the arc process; these gases eliminate the need for a flux and enable formation of a much stronger joint.

Weld efficiencies as high as 95 percent are common when good magnesium welding techniques are used. However, because of the high thermal expansion quality of the metal, stresses caused by welding are often severe enough to cause fractures, and the welded parts should be stress-relieved to minimize that possibility.

Vibration Damping

The vibration resistance and shock absorption property of magnesium is more efficient than that of any metal other than certain aluminum alloys. For this reason, magnesium plate was selected for the vibration testing fixture shown in Fig. 8–2.

Fig. 8–2. Magnesium vibration testing fixture. (*Courtesy, The Dow Chemical Company*)

Other Properties

Many other physical, chemical, and mechanical properties make magnesium alloys extremely useful as materials for jigs and fixtures in many special applications and under special conditions. These qualities should be studied closely to determine their applicability to a special tooling problem.

Chemical Stability. Alloys of magnesium remain unaffected by coolants, lubricants, and other solutions that customarily come into contact with jigs and fixtures. The metal forms an oxide coating after prolonged exposure to air, and this coating will protect the tool indefinitely under normal conditions. For added protection, or for storage indoors, a coating of light oil (SAE 10) may be applied to the tool; for out-of-doors use or storage the magnesium jig or fixture should be painted just as any other metal.

Low Friction. The low coefficient of friction and the nongalling properties of magnesium can be used to advantage whenever a bearing or sliding surface against another material is needed. Magnesium jigs and fixtures can be designed to make use of this property in rapid loading and unloading features.

Nonsparking Properties. Although pure copper is the only metal that will not spark, magnesium will discharge sparks only under very special conditions of high temperature and high machining speeds.

Nonmagnetic Properties. Magnesium is virtually nonmagnetic. Like aluminum, it cannot be permanently affected by a magnetic field. This property may provide advantages when magnetic machine tables are used or when magnetic alloys are being machined.

MAGNESIUM FIRE HAZARD

Although magnesium has been considered a hazardous metal to machine because of its flammability, the danger has perhaps been exaggerated. Experience has shown that it can be machined with little difficulty when properly handled and when proper safety precautions are followed.

Flammability

Magnesium must melt before it will ignite. Depending on the alloy content, the metal will begin melting at temperatures from 800° F to 1200° F. Because of the low coefficient of friction and the high heat conductivity of magnesium, the metal will seldom attain such temperatures in rough machining operations, since it will dissipate frictional heat almost as fast as the heat is generated. For this reason, most magnesium machining operations can be performed dry. Roughing and semifinishing cuts seldom need a coolant even at extremely high speeds as long as sharp tools are used.

In finish-machining operations the fire hazard is more serious, for the fine chips and dust produced are highly combustible, and more frictional heat is generated. The solution to this problem is to use a cutting fluid or coolant to conduct heat away from the cutting area and prevent sparks. Coolants may also be used for work that requires extra-close tolerance to help maintain uniform temperatures and reduce distortion in the part.

Commercial cutting fluids have been developed for machining magnesium, and it is best to use one of these. If a commercial fluid is for some reason not readily available, a mixture of kerosene and mineral oil, or straight mineral oil, will serve satisfactorily.

CAUTION: Never use a pure solution of animal or vegetable oil for machining magnesium. Never use water-soluble oils, oil-water emulsions, or water solutions of any kind. These solutions, especially the water-mixture coolants, are dangerous, for they can greatly intensify any chip fire that may occur.

Fire Protection

Because of the fire hazard of magnesium machining, everyone working with the metal should fully understand fire precautions and proper extinguishing procedures.

A magnesium chip fire usually burns with a spectacular soft white glow, but it is harmless when properly controlled. Magnesium fires are best extinguished by smothering compounds such as commercial extinguishing powders, graphite-based powders and compounds, or dry cast-iron chips. A container of one of these materials, together with a shovel, should be placed near all areas where magnesium is being machined. *CAUTION: Water-moist materials or standard liquid fire extinguishers should never be used on a magnesium fire.* Water will enter into a chemical reaction with burning magnesium, generating hydrogen and causing eruptions of the flames.

In the initial stages a magnesium fire should be isolated by scraping away the chips adjacent to it. The smothering material should then be shoveled gently over the burning area to avoid scattering the fire. If the fire is on a nonflammable surface such as a concrete floor or machine-tool surface, it is only necessary to cover the flame well and to stir the smothering material into areas that continue to smoke.

If the fire is on a wooden floor or other flammable surface, it should be covered with a smothering compound, then gently lifted with a shovel and placed in a metal or asbestos container. If this procedure is not feasible, the smothering compound should be mixed well into the base of the fire to break contact between the fire and the flammable surface.

To reduce the risk of fire, magnesium chips and dust should not be permitted to collect on, under, or around machine tools, or any other place for that matter. Chips should be removed frequently and stored in tightly covered, nonflammable containers.

Because of the natural surface finish of magnesium, grinding or polishing, even after rough machining, will be necessary only when extreme accuracy is required. Both these operations are more hazardous than rough machining, but fires can be avoided with proper precautions. Equipment used for grinding or polishing magnesium must not be used for ferrous or sparking materials until the magnesium dust has been thoroughly removed. Grinding wheels and sanding belts should be labeled "For Magnesium Only."

No smoking should be allowed in the grinding or polishing areas, and aprons of tightly woven fabric should be worn by all workers in the area to minimize the accumulation of magnesium dust on their clothing. Cloth containing magnesium dust is highly combustible, and coarse-textured or fuzzy cloth will accumulate the dust rapidly. As an added precaution, protective clothing and aprons should be brushed frequently and changed daily.

MAGNESIUM STOCK FOR TOOLING

Magnesium, in the form of tooling plate and extrusions, has replaced hot- and cold-rolled steel for a wide range of tooling applications. Not only is the lightweight material economical for jig and fixture construction, but its savings continue during the tool's use because of reduced operator fatigue.

In the aircraft and missile industries magnesium tooling is widely used since its coefficient of thermal expansion is approximately the same as that of other aircraft parts and it consequently offers greater interchangeability among the parts manufactured. Engineers at one aircraft plant credited magnesium tooling with saving their company $93,000 in one year—not to mention the savings in production areas where less manpower was required and worker fatigue was greatly reduced.*

---

*Jack M. Hockett, "Putting Magnesium to Work," *The Tool and Manufacturing Engineer*, July, 1961.

Tooling Plate

Magnesium plate suitable as tooling material is presently available in standard thicknesses from .250 to 3.500 in. Mill standard sizes are 48 × 96 in, 44 × 144 in, 60 × 144 in, and 72 × 144 in, depending on the gage. Table VIII–2 gives the available standard thickness and sizes.

Table VIII–2. Mill Standard Sizes of Magnesium Tooling Plate.

Thickness (in)	Size (in)			
.250, .375	48 × 96	48 × 144	—	—
.500, .625, .750,				
1.000	48 × 96	48 × 144	60 × 144	72 × 144
1.250, 1.500, 1.750	48 × 96	48 × 144	60 × 144	72 × 144
2.000, 2.500, 3.000	48 × 96	48 × 144	60 × 144	72 × 144
3.500	48 × 96	—	—	—

(Courtesy, The Dow Chemical Company)

Typical mechanical properties of all gages of magnesium tooling plate are as follows:

1) Tensile strength — 35,000 psi
2) Tensile yield strength — 19,000 psi
3) Compressive yield strength — 10,000 psi
4) Elongation — 10 percent in 2 in.

Accuracy. Magnesium tooling plate is manufactured under very closely controlled conditions to provide the flatness necessary for tooling purposes. Careful rolling procedures and controlled thermal flattening yield a product with flatness tolerances and surface finish (60 to 70 microinches) satisfactory for about 90 percent of all jig and fixture requirements. Table VIII–3 gives commercial flatness tolerances currently available. For most practical applications, these surfaces would require no additional costly machining.

Table VIII–3. Flatness Tolerances of Magnesium Tooling Plate.

Thickness (in)	In Any One Foot		In Any Six Feet	
	Guaranteed	Typical	Guaranteed	Typical
.250–1.000	.005	.003	.015	.007
1.001–3.500	.010	.005	.020	.010

(Courtesy, The Dow Chemical Company)

Because of the annealing and rolling process magnesium tooling plate is usually given, all residual stresses are eliminated. No stress relief is required after machining, and changes in temperature will not cause warpage, so the accuracy of the machining will be retained even under varying environmental conditions.

When accuracy is required, as for gaging fixtures, magnesium tooling plate is ideal. An excellent example of such a fixture is shown in Fig. 8–3. It was cut

Fig. 8–3. Magnesium checking fixture. (*Courtesy, The Dow Chemical Company*)

from tooling plate with a band saw, then clamped together and arc welded. The three gaging surfaces were then machined, drilled, and reamed for inserts. Because of the magnesium plate's stress stability, the surfaces remained parallel within .001 in. even after months of continued use.

Cost. Magnesium tooling plate is considerably more expensive per pound than other tooling material, but because of its lower density, price comparisons should be made on a basis of volume rather than weight. By volume, magnesium plate costs an average of 22 percent less than other tooling metals with comparable surface tolerances. As an example, compare the costs of the 1.5 in × 48 in × 60 in. materials listed in Table VIII–4. Note that the magnesium tooling plate costs much less than the type 200 and 300 aluminum plates. In addition, there is an extra charge for cutting the aluminum. Although the price of rough boiler plate is about one-half the price of the magnesium, the added cutting and grinding costs raise the boiler plate's costs to a range where magnesium is competitive.

Table VIII–4. Tooling Plate Comparative Costs (1.5 in × 48 in × 60 in Plate).

Material	Weight (lbs)	Price	Extra Cutting Charge	Extra Grinding Charge	Total Price To User
Boiler plate	1242	$11.07/cwt ($137.49)	$5.49 (burning)	$57.60 (± .010) (grinding 2 sides)	$200.58
Aluminum (type 200)	424	$75.80/cwt ($321.49)	$3.00 (sawing)	—	$324.49
Aluminum (type 300)	424	$87.40/cwt ($370.58)	$3.00 (sawing)	—	$373.58
Magnesium (B&P)	276	$.73 + $.27/lb ($276.00)	—	—	$276.00

(Courtesy, Brooks & Perkins, Inc.)

Even though steel plate costs less than magnesium, material cost is only one part of total tool cost, and the better machinability of magnesium more than makes up the difference. To better illustrate this point, Table VIII–5 shows detail estimates of the same jig using three different materials. Each operation is estimated individually, and a labor rate of $10.00/hr is used. Note that the magnesium jig would take the least time and money to machine.

Table VIII–5. Cost Comparison of Three Jig Materials.

| | Tool Plate | | | | | |
| | Aluminum | | Magnesium | | Hot Rolled Steel | |
Operation	Hours	Dollars	Hours	Dollars	Hours	Dollars
Obtain stock and shear to size	.2	2.00	.2	2.00	.3	3.00
Face top and bottom sides of base plate. Rough grind on Blanchard #18 surface grinder.	Not required		Not required			
Setup					.4	4.00
Facing					.3	3.00
Layout complete.	2.5	25.00	2.5	25.00	2.5	25.00
Shape all edges, break corners. Use 32 in American shaper M883.						
Setup	.2	2.00	.2	2.00	.2	2.00
Face, File	1.3	13.00	1.1	11.00	1.9	19.00
Mill notch on one edge $^3/_{16} \times 1^1/_4 \times ^1/_2$. Use Mil. milling machine M131.						
Setup	.3	3.00	.3	3.00	.3	3.00
Mill	.21	2.10	.18	1.80	.3	3.00
Mill holes for 53-drill bushings, two holes for locating pins, and four holes for lifting handles. Use Mil. milling machine M131.						
Setup	.3	3.00	.3	3.00	.3	3.00
Boring	2.2	22.00	1.9	19.00	3.1	31.00
Turn, point, and part off two locating pins. Use lathe M108.						
Setup	.2	2.00	.2	2.00	.2	2.00
Lathe	.3	3.00	.3	3.00	.3	3.00
Hand tap four blind holes for lifting handles capscrews.						
Setup	.2	2.00	.2	2.00	.2	2.00
Tap	.3	3.00	.24	2.40	.4	4.00

Table VIII–5. Cost Comparison of Three Jig Materials *(continued).*

| | Tool Plate | | | | | |
| | Aluminum | | Magnesium | | Hot Rolled Steel | |
Operation	Hours	Dollars	Hours	Dollars	Hours	Dollars
Obtain stock and shear to size	.2	2.00	.2	2.00	.3	3.00
Press 53-drillbushings, two locating pins into base.						
Setup	.20	2.00	.2	2.00	.2	2.00
Press	1.5	15.00	1.5	15.00	1.5	15.00
Assemble lifting handles to base plate with four capscrews.						
Job	.3	3.00	.3	3.00	.3	3.00
Locate and stamp jig with 27 characters	.2	2.00	.2	2.00	.25	2.50
Labor machining totals*	10.41	104.10	9.82	98.20	12.95	129.50

*Based on labor rate of $10.00/hr. Hole locating machine could be used on this job which would result in a savings of $25.00 on all three types of material.
(Courtesy, The Dow Chemical Company)

A similar detail estimate, Table VIII–6, compares the total costs of a checking fixture made from the three metals. This estimate includes plate and other material costs as well as machining costs. The resulting savings of the magnesium fixture is $56.36 over the aluminum fixture and $344.01 over the mild steel. Also note that the majority of these savings are in the area of machining.

Table VIII–6. Cost Comparison of Aluminum, Magnesium, and Mild Steel Fixtures.

| | | | Tool Plate | | | | | |
| | | | Alum. | | Mag. | | Mild Steel | |
Detail No.	No. of Pcs.	Operations	Hours	Dollars	Hours	Dollars	Hours	Dollars
1 thru 25	44	Engine Lathe M102						
		Turn Point and Part Off	5.3	53.00	5.3	53.00	5.3	53.00
		Setup	.2	2.00	.2	2.00	.2	2.00
		Mil. Mill M113						
		Mill Flats	9.0	90.00	9.0	90.00	9.0	90.00
		Setup	.4	4.00	.4	4.00	.4	4.00
26	1	32 in. American Shaper M112						
		Shape 4 Edges	.84	8.40	.72	7.20	1.2	12.00
		Shape 2 Edges	Not required		Not required		.8	8.00
		Setup	.2	2.00	.2	2.00	.2	2.00
		Layout Holes	.5	5.00	.5	5.00	.5	5.00

Table VIII–6. Cost Comparison of Aluminum, Magnesium, and Mild Steel Fixtures (*continued*).

Detail No.	No. of Pcs.	Operations	Alum.		Mag.		Mild Steel	
			Hours	Dollars	Hours	Dollars	Hours	Dollars
		Mil. Milling Machine M132						
		Mill Ream Holes	1.75	17.50	1.5	15.00	2.5	25.00
		Setup	.3	3.00	.3	3.00	.3	3.00
		Hand Tap 1 Hole	.07	.70	.05	.50	.1	1.00
		Setup	.2	2.00	.2	2.00	.2	2.00
		Hand Stamp Tolerance Marks	.4	4.00	.4	4.00	.54	5.40
27	2	32 in. American Shaper M112						
		Shape 4 Edges	1.54	15.40	1.32	13.20	2.2	22.00
		Shape 2 Sides	Not required		Not required		1.6	16.00
		Setup	.2	2.00	.2	2.00	.2	2.00
		Layout Holes	1.0	10.00	1.0	10.00	1.0	10.00
		Mil. Milling Machine M132						
		Mill Ream all Holes	3.5	35.00	3.0	30.00	5.0	50.00
		Setup	.3	3.00	.3	3.00	.3	3.00
		Hand Tap 1 Hole in Each Pc.	.14	1.40	.1	1.00	.2	2.00
		Setup	.2	2.00	.2	2.00	.2	2.00
		Hand Stamp Tolerance Marks	.8	8.00	.8	8.00	1.08	10.00
28	2	32 in. American Shaper M112						
		Shape 4 Edges	1.47	14.70	1.26	12.60	2.1	21.00
		Shape Sides	Not required		Not required		1.5	15.00
		Setup	.2	2.00	.2	2.00	.2	2.00
		Layout Holes	.9	9.00	.9	9.00	.9	9.00
		Mil. Milling Machine M132						
		Mill Ream Holes	2.94	29.40	2.52	25.20	4.2	42.00
		Setup	.3	3.00	.3	3.00	.3	3.00
		Hand Tap 1 Hole in Each Pc.	.14	1.40	.1	1.00	.2	2.00
		Hand Stamp Tolerance Marks	.67	6.70	.67	6.70	.9	9.00
29	2	32 in. American Shaper M112						
		Shape 4 Edges	1.35	13.50	1.14	11.40	1.9	19.00
		Shape 2 Sides	Not required		Not required		.7	7.00
		Setup	.2	2.00	.2	2.00	.2	2.00
		Layout Holes	.8	8.00	.8	8.00	.8	8.00
		Mil. Milling Machine M132						
		Mill Ream Holes	2.4	24.00	2.0	20.00	3.4	34.00
		Setup	.3	3.00	.3	3.00	.3	3.00
		Hand Tap	.14	1.40	.1	1.00	.2	2.00
		Setup	.2	2.00	.2	2.00	.2	2.00
		Hand Stamp Tolerance Marks	.53	5.30	.53	5.30	.73	7.30
30	2	32 in. American Shaper M112						
		Shape 4 Edges	1.2	12.00	1.0	10.00	1.7	17.00
		Shape 2 Sides	Not required		Not required		.6	6.00
		Setup	.2	2.00	.2	2.00	.2	2.00
		Layout Holes	.8	8.00	.8	8.00	.8	8.00

Table VIII–6. Cost Comparison of Aluminum, Magnesium, and Mild Steel Fixtures (*continued*).

Detail No.	No. of Pcs.	Operations	Alum. Hours	Alum. Dollars	Mag. Hours	Mag. Dollars	Mild Steel Hours	Mild Steel Dollars
		Mil. Milling Machine M132						
		Mill Ream Holes	2.4	24.00	2.0	20.00	3.4	34.00
		Setup	.3	3.00	.3	3.00	.3	3.00
		Hand Tap Holes	.14	1.40	.1	1.00	.2	2.00
		Setup	.2	2.00	.2	2.00	.2	2.00
		Hand Stamp Tolerance Marks	.53	5.30	.53	5.30	.73	7.30
31	2	32 in. American Shaper M112						
		Shape 4 Edges	1.9	19.00	1.6	16.00	2.7	27.00
		Shape 2 Sides	Not required		Not required		1.8	18.00
		Setup	.2	2.00	.2	2.00	.2	2.00
		Layout Holes	.3	3.00	.3	3.00	.3	3.00
		Drill and Tap (M98)	1.05	10.50	.9	9.00	1.5	15.00
		Setup	.3	3.00	.3	3.00	.3	3.00
32	1	32 in. American Shaper M112						
		Shape 4 Edges and 1 Slot	1.05	10.50	.9	9.00	1.5	15.00
		Shape 2 Sides	Not required		Not required		1.2	12.00
		Setup	.2	2.00	.2	2.00	.2	2.00
		Layout	.6	6.00	.6	6.00	.6	6.00
		Mil. Milling Machine M132						
		Mill and Co'bore Holes	1.0	10.00	.84	8.40	1.4	14.00
		Setup	.3	3.00	.3	3.00	.3	3.00
		Press in 2 Drill Bushings	.1	1.00	.1	1.00	.1	1.00
		Setup	.2	2.00	.2	2.00	.2	2.00
33	1	32 in. American Shaper M112						
		Shape 4 Edges	.9	9.00	.8	8.00	1.3	13.00
		Shape 2 Sides	Not required		Not required		1.2	12.00
		Setup	.2	2.00	.2	2.00	.2	2.00
		Layout	.7	7.00	.7	7.00	.7	7.00
		Mil. Milling Machine M132						
		Mill and Co'bore Holes	1.3	13.00	1.1	11.00	1.8	18.00
		Setup	.3	3.00	.3	3.00	.3	3.00
		Press 4 Drill Bushings	.2	2.00	.2	2.00	.2	2.00
		Setup	.2	2.00	.2	2.00	.2	2.00
39	2	Engine Lathe M102						
		Turn, Chamfer, and Part Off	.3	3.00	.3	3.00	.3	3.00
		Setup	.2	2.00	.2	2.00	.2	2.00
40	1	Engine Lathe M102						
		Turn, Chamfer, and Part Off	.2	2.00	.2	2.00	.2	2.00
		Setup	.2	2.00	.2	2.00	.2	2.00
	7	Subassemble Details						
		# 31, 32, 33, 39, 40.	2.0	20.00	2.0	20.00	2.5	25.00
			2.0	20.00	2.0	20.00	2.2	22.00

Table VIII–6. Cost Comparison of Aluminum, Magnesium, and Mild Steel Fixtures (continued).

Detail No.	No. of Pcs.	Operations	Tool Plate					
			Alum.		Mag.		Mild Steel	
			Hours	Dollars	Hours	Dollars	Hours	Dollars
38	1	Layout Complete						
		Horizontal Mill M142						
		Face Top, Bottom Sides	Not required	Not required	Not required	Not required	3.4	34.00
		Setup	Not required	Not required	Not required	Not required	.2	2.00
		32 in. American Shaper M112						
		Shape 4 Edges	1.4	14.00	1.2	12.00	2.0	20.00
		Setup	.2	2.00	.2	2.00	.2	2.00
		Mil. Milling Machine M132						
		Mill and Co'bore Holes	2.1	21.00	1.8	18.00	3.0	30.00
		Setup	.3	3.00	.3	3.00	.3	3.00
		Hand Tap 5 Holes	.5	5.00	.4	4.00	.6	6.00
		Setup	.2	2.00	.2	2.00	.2	2.00
34	4	Engine Lathe M102						
		Turn, Knurl, and Part Off						
		Drill and Tap	1.6	16.00	1.6	16.00	1.6	16.00
		Setup	.2	2.00	.2	2.00	.2	2.00
49	2	32 in. American Shaper M112						
		Shape Complete	1.6	16.00	1.6	16.00	1.6	16.00
		Setup	.2	2.00	.2	2.00	.2	2.00
35	1	Engine Lathe M102						
		Chamfer and Thread						
		Both Ends	.2	2.00	.2	2.00	.2	2.00
		Setup	.2	2.00	.2	2.00	.2	2.00
36	4	Engine Lathe M102						
		Part Off and Face Both Ends	.4	4.00	.4	4.00	.4	4.00
		Setup	.2	2.00	.2	2.00	.2	2.00
37	2	Engine Lathe M102						
		Chamfer and Thread						
		Both Ends	.4	4.00	.4	4.00	.4	4.00
		Setup	.2	2.00	.2	2.00	.2	2.00
42	3	Engine Lathe M102						
		Turn, Knurl, and Part Off						
		Chamfer and Thread						
		One End	1.8	18.00	1.8	18.00	1.8	18.00
		Setup	.2	2.00	.2	2.00	.2	2.00
45	2	Engine Lathe M102						
		Part Off and Chamfer	.1	1.00	.1	1.00	.1	1.00
		Setup	.2	2.00	.2	2.00	.2	2.00
46	1	32 in. American Shaper M112						
		Shape 4 Edges	.6	6.00	.6	6.00	.6	6.00

Table VIII–6. Cost Comparison of Aluminum, Magnesium, and Mild Steel Fixtures *(continued)*.

Detail No.	No. of Pcs.	Operations	Tool Plate				Mild Steel	
			Alum.		Mag.		Mild Steel	
			Hours	Dollars	Hours	Dollars	Hours	Dollars
		Setup	.2	2.00	.2	2.00	.2	2.00
		Layout Complete	.2	2.00	.2	2.00	.2	2.00
		Mil. Milling Machine M132						
		Mill and Countersink Holes	.3	3.00	.3	3.00	.3	3.00
		Setup	.3	3.00	.3	3.00	.3	3.00
		Final Assembly Stamp tool number on all loose parts. Assemble and ream where necessary. Check for operating condition. Prepare to ship.	13.0	130.00	13.0	130.00	14.0	140.00
		Get Out Stock Material	3.0	30.00	2.7	27.00	4.0	40.00
		Subtotals (Labor Machining)	90.85	908.50	85.78	857.80	120.68	1206.80

	Aluminum Tooling Plate	Magnesium Tooling Plate	Hot Rolled Steel
Labor Machining Costs*	$908.50	$857.80	$1206.80
Plate Cost**	23.92	18.26	13.27
Material Cost (excluding plate)	22.00	22.00	22.00
Total	$954.42	$898.06	$1242.07

* Based on labor rate of $10.00/hr.
** Based on purchase of material at base price (Mg & Al = 30,000 lbs; steel = order quantity 20,000 lbs or over, item quantity 10,000 lbs or over). Shearing and sawing charges are incorporated in labor machining costs.

Plate material used:		Al (lbs)	Mg (lbs)		Steel	(lbs)
1 pc.	.5 × 3 × 22.5 in	3.25	2		.5 × 3 × 11.5 in	5
1 pc.	.5 × 4.25 × 11 in	2.25	1.5		.625 × 3.5 × 11 in	7
1 pc.	.75 × 1.25 × 76.25 in	7	4.5		.625 × 5 × 11.5 in	10
1 pc.	.75 × 11.5 × 30.75 in	26	17		1 × 1.25 × 76.25 in	27
		39.5	25		1 × 11.5 × 30.75 in	101
						150

(Courtesy, The Dow Chemical Company)

Extrusions

Magnesium is available commercially in a wide variety of extrusions. These forms, some of which are shown in Fig. 8–4, include rods, bars, tubes, angles, V, I, T, and Z sections, and a variety of channeled shapes. Each of these is usually available in 12-ft lengths.

Rods are available in sizes from .25 in. to 5 in. in diameter; bars come in square — .25 in. to 4 in; hexagonal — .25 in. to 4 in. across flats; and rectangular — .125 in. to 3.5 in. thick, .1875 in. to 11 in. wide. Tubing ranges from .25-in to 12-in diameters and comes in various wall thicknesses.

Fig. 8–4. Magnesium extrusions. (*Courtesy, The Dow Chemical Company*)

Fig. 8–5. Locating fixture made with square magnesium tubing. (*Courtesy, The Dow Chemical Company*)

The fixture shown in Fig. 8–5 illustrates the use of magnesium extrusions. It was constructed to locate details on an external aircraft fuel tank, and magnesium square tubing was chosen for the frame because of its light weight and inherent strength.

9

wood tool construction

Wood is man's oldest and most readily available structural material. Wood is inexpensive, lightweight, and can be readily shaped or formed. In addition, it is a nonstrategic material which can take the place of more vital materials when such materials are in short supply. For these reasons, wood is a favorable construction material for low-cost jigs and fixtures.

In most cases, wood can substitute for more expensive average-duty steel jigs and fixtures. It can also be used advantageously for large structural tooling when material represents the largest proportion of the tool's cost. However, it should be clearly understood at once that wood tools should not be used for extremely close work or when tooling is expected to be heavily used or abused. In this chapter we examine the advantages, disadvantages, and characteristics of natural and processed wood and the design, fabrication, and economics of wood tools.

ADVANTAGES AND DISADVANTAGES OF WOOD FOR TOOLING

Wood has a number of relevant properties, some favorable and some unfavorable. These should be thoroughly understood before the decision is made to use wood for any given jig or fixture application. The following favorable properties should be checked for their applicability to a given tooling problem:

1) Wood can be formed into various shapes with simple tools and relatively limited skill.
2) Because of its porosity and cellular nature, wood is one of the few major raw materials that can be easily fastened with nails, screws, or glue.
3) The compression strength of wood is very high. In fact, when the wood is free of voids, its strength is higher in proportion to its weight than steel's. Wood also has moderate shear strength.
4) The expansion and contraction of wood due to temperature changes is negligible. Its rate of thermal expansion is less than one-half that of steel.
5) Wood absorbs shocks and vibrations better than most other construction materials.
6) Wood does not oxidize and will resist the action of acids and salts in normal use.

On the other side of the ledger, wood has a number of properties which might be considered unfavorable for certain applications. The tool engineer should seriously consider the following before selecting wood for his tools:

1) The principal disadvantage of wood is its tendency to warp, shrink, or swell with marked changes in humidity.
2) Wood, being an organic substance, cannot be subjected to extreme heat, nor can it be used in areas where fire hazards exist. Chemically treated fire-resistant wood is now available, however, at only about 5 percent more in price.
3) The hardness of wood is limited and there is no commercially successful method by which its hardness can be materially increased without changing its character.
4) Most woods split easily parallel to the grain.
5) Wood cannot be extruded or rolled into new shapes.

In some cases these disadvantages can be corrected or minimized. To use wood advantageously for jig and fixture construction, a knowledge of this material's structure and physical characteristics is necessary. This chapter provides a short explanation of the fundamentals of woodworking to assist tool engineers in using wood for low-cost, efficient tooling.

CHARACTERISTICS OF WOOD

Wood is divided into two general classes—hardwood and softwood. These terms are imprecise, for no definite degree of hardness divides hardwoods from softwoods. The terms are used in a general sense to indicate the type of tree—whether deciduous or evergreen—from which the wood is taken. Therefore, when wood is selected its actual rated strength and not the term by which it is known should be considered. A realistic classification of the properties of the most common species of trees used for construction is listed in Table IX-1.

Table IX-1. Properties of Wood.

Wood	Strength	Shrinkage	Workability	Resistance to Shock	Resistance to Splitting
SOFTWOOD					
Cedar	Medium	Very Little	Excellent	Excellent	Good
Cypress	Medium	Moderate	Poor	Poor	Good
Fir, Douglas	Medium	Moderate	Poor	Poor	Poor
Pine, White	Low	Very Little	Excellent	Poor	Good
Spruce	Low	Moderate	Fair	Poor	Good
HARDWOOD					
Ash	High	Moderate	Good	Good	Fair
Beech	High	Considerable	Poor	Good	Fair
Birch	High	Considerable	Fair	Excellent	Fair
Cherry	High	Moderate	Fair	Good	Fair
Chestnut	Medium	Moderate	Good	Fair	Poor
Elm	Medium	Considerable	Poor	Excellent	Excellent
Maple, Hard	High	Moderate	Good	Good	Fair
Hickory	High	Considerable	Poor	Excellent	Poor
Oak	High	Considerable	Fair	Good	Poor
Walnut	High	Moderate	Good	Good	Fair

Growth Rings

The cross-section of a log from any type of tree will show many concentric layers or rings as illustrated in Fig. 9–1. These layers are made up of millions of cells of wood fiber and are commonly known as growth rings, or annual rings, since a new layer appears each year that the tree continues to grow. The growth rings differ greatly in width and density among different species and among

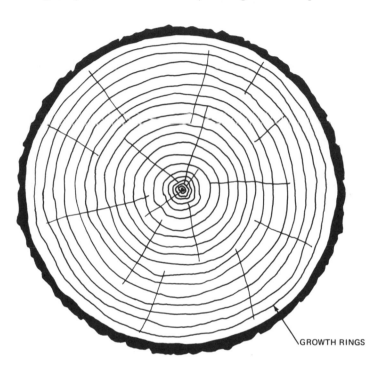

Fig. 9–1. Cross-section of a tree showing the growth rings.

trees of the same species. They also differ at various heights within a given tree and with the amount of rainfall that the tree has received from year to year. Generally, however, fast-growing trees develop wider growth rings than slow-growing trees.

Strength

The quality of wood, particularly its density and strength, is directly related to the growth rings in all tree species. Wide-ringed wood generally indicates lumber of inferior quality. Not only is this type of wood weaker than normal wood, but it is also subject to high degrees of shrinkage and warpage. For this reason, rapidly grown trees are less likely to provide satisfactory wood than trees whose growth rates are slower. Wood of the best quality for tooling construction is usually produced when the growth rate results in more than six growth rings per inch.

The direction of the force or load in relation to the growth rings is also important to wood's maximum strength. Usually very little attention is paid this factor, but experiments and tests have proven that greatest strength is obtained when the growth rings lie perpendicular to the load. When the rings lie parallel to the load, medium strength is obtained, but when the growth rings are at 45 deg to the load the strength of the wood is considerably reduced. These positions of the growth rings in relation to the force or load applied to the wood are shown in Fig. 9–2.

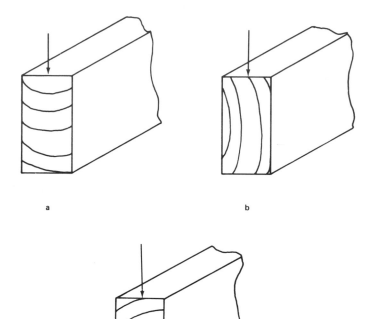

Fig. 9–2. Forces or loads in relation to the growth rings in wood — *a*, perpendicular; *b*, parallel; *c*, at 45 deg.

Warpage

The main disadvantage of wood is its tendency to shrink or swell with changes in humidity. When such shrinkages or swellings occur across areas of different density in the wood, warpage may result. *Warpage* is the change or variation from a flat or plane surface of a board.

The tendency for a board to warp is usually a result of the difference in the density between new and old cells in different growth layers of the wood. A look at Fig. 9–3 will show this relationship. Side A of the board is closer to the heart of the tree, as indicated by the growth rings, and contains older wood than side B. As the board dries, the younger and more open cells on side B shrink more

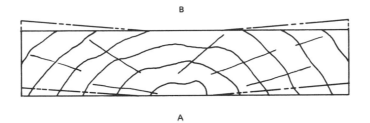

Fig. 9–3. Direction of warpage of a drying board.

rapidly and to a greater extent than the denser cells on side A, therefore causing the board to bend or warp in the direction indicated by the dotted line. If the board becomes moist, the cells on side B swell more rapidly than those on side A and cause the board to bend in the opposite direction. Boards therefore warp according to their position in the tree from which they were cut. Fig. 9–4 indicates how their position and variation in growth rings affect warpage.

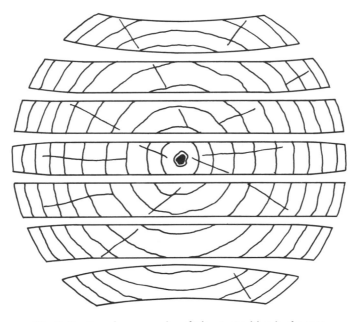

Fig. 9–4. Board warpage in relation to position in the tree.

The warpage of boards can be countered to some extent by laminated construction in which two or more boards are bonded together with an adhesive. For best results, it is necessary to glue the boards so either their outer faces or inner faces are together. This process is shown in Fig. 9–5a. For three-piece laminated construction, as illustrated in Fig. 9–5b, the two outside boards should have their inner faces exposed. Fig. 9–5c shows a four-layer construction, with the inner faces of the two center boards together and the inner faces of the outer boards exposed.

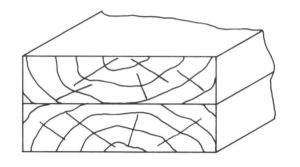

a

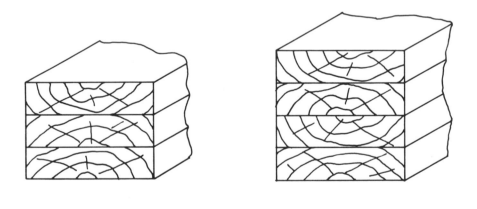

b   c

Fig. 9–5. Laminated construction to offset warpage—a, two layers; b, odd number of layers; c, even number of layers.

PROCESSED WOOD

Processed wood is wood that has been re-formed or altered by some type of manufacturing process. By the processes which the wood undergoes, some of its unfavorable properties—warpage, splitting, and limited hardness—are reduced, while some of its favorable properties—strength, limited expansion and contraction, and shock absorption — are improved. Three types of processed wood are especially useful for tool construction. They are plywood, metal-clad plywood, and compressed wood.

Plywood

Plywood is wood paneling composed of wood layers bonded together with moisture-resistant or moisture-proof glue. The layers are juxtapositioned so that their grains run in different directions, generally at right angles to each other, so the resultant plywood sheet is equally strong in all directions. There are two main types of plywood, exterior and interior; the difference is in the glue used in their construction. Exterior plywood, bonded with moisture-proof glue, is best for tooling construction because it resists the processing coolants and oils better than interior plywood.

Plywood panels are commercially available in thicknesses ranging from $\frac{1}{8}$ in. to $1\frac{1}{4}$ in. Panels are available in almost any desired size, but 48×96 in. is the most popular. Larger panels are available, but the price per square foot is higher. Plywood is also available in formed angles of various sizes and in tubing ranging in diameter from 2 in. to 16 in.

In addition to its added strength and resistance to warping, splitting, and checking, plywood will not expand or contract to the extent of regular wood because of its cross-grained construction. The "working" or expansion and contraction of plywood under load is reduced by the bonding process to less than the normal movement of wood along the grain. Actually, a simple three-ply plywood will expand less than .001 in/in even when exposed to dampness.

For increased versatility as a tooling construction material, exterior panels of plywood can be soaked with water or steamed and bent to almost any curvature. Although most commercial plywood panels are intended for flat use, they can be more easily bent or formed than solid wood of equivalent thickness.

Metal-Clad Plywood

Metal-clad (or metal-faced) plywood is fabricated by bonding thin sheets of metal to one or both faces of a piece of plywood. The resulting product is equivalent in rigidity to a solid steel section four times its weight. Many different types of metal are used to face the panels, but the most common are galvanized steel, aluminum, stainless steel, and Monel* metal. Stock panels are generally available in widths of 30 in. to 40 in. and lengths of 144 in.

The advantages of metal-clad plywood as a tooling material are as follows:

1) *Good resistance to buckling and bending.* — Metal-clad plywood offers greater resistance to buckling and bending, per unit weight, than any other construction material.
2) *Good wear resistance.* — Wear resistance characteristics are created by the metal facing.
3) *Resistance to damage by surface blows.* — Unless struck by sharp objects, this type of plywood is not vulnerable to damage from heavy use. The plywood core has a tendency to absorb the energy of a blow and distribute

---

*Reg. TM, International Nickel Company, Inc.

it over a large surface instead of buckling or breaking as a thin sheet of metal would do.

4) *Resistance to heat and fire.* — Metal-clad plywood, as compared with unclad plywood, has substantial advantages in this area.

5) *Ease in working.* — It can be cut easily with standard wood or metal-cutting tools.

In addition to the characteristics listed above, metal-clad plywood possesses certain other characteristics useful to the toolmaker. For example, the edges of the plywood may be protected by cutting back the plywood core and folding over either one or both of the exposed metal edges. This is illustrated in Fig. 9–6. The assembly of tools constructed with this type of plywood is simplified

SOLDER

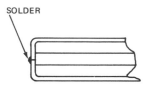

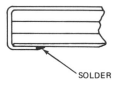

SOLDER

DOUBLE-CLAD

SINGLE-CLAD

Fig. 9–6. Methods of protecting exposed edges of metal-clad plywood.

also since the plywood can be fastened with nails or screws in the same way as any other type of wood material.

Finally, this material is easy to shape. Plywood with a metal facing on one side only may be bent or formed to create minor curves. In this process the metal must be on the outside of the curve. The material must be processed through bending rolls, as with sheet metal, to obtain more acute curves.

Compressed Wood

Compressed wood is a manufactured product composed of many layers or veneers of wood ranging in thickness from .01 to .08 in. and bonded with phenolic resin under high pressure and heat. The amount of pressure applied determines the density of the finished product. By varying the conditions of pressure and temperature, manufacturers can produce compressed wood in a wide range of hardnesses.

Because of its favorable characteristics, compressed wood is being used more and more for the construction of jigs and fixtures for limited production. At present its two main applications are in the aircraft industry and in the fabrication of steel-rule dies.

Its main advantage is its excellent machinability. Compressed wood can be machined easily with standard woodworking tools at speeds and feeds comparable to those recommended for natural wood. These excellent machinability characteristics produce man-hour-tool economies not possible with conventional materials.

Compressed wood is commercially available in standard board sizes up to 6 in. thick. In 36 × 84-in board sizes, it is available in thicknesses ranging from .25 to 6 in. Greater thicknesses may be obtained by joining several sheets or boards with cold resin glue. The usual tensile strength of compressed board is 60,000 psi, and its usual compressive strength is 40,000 psi.

DESIGNING AND FABRICATING WOOD TOOLS

The design of wood jigs and fixtures is basically the same as the design of any equivalent conventional tool. However, the designer must keep in mind the unusual conditions that wood fabrication presents. To meet the service requirements and to avoid distortion of the tool through warping and fabrication stresses, the designer must:

1) Select the proper wood for the tool
2) Be familiar with basic wood construction methods, joinery, and hardware
3) Protect the wood properly after assembling the tool.

It is often preferable to build wood tools in the pattern or woodworking shop. Toolmakers and machinists, whose skills normally are in the area of metalworking, may not possess the necessary knowledge and capabilities to work with wood, and in addition they may not have the correct shop tools. Patternmakers, on the other hand, have the knowledge, skill, and abilities to handle woodworking, and they can be efficient and economical in producing wood tooling. However, in order to construct and add the metal details that are often used on wood tools, the patternmaker should have a practical knowledge of metalworking techniques and should be equipped with basic metalworking tools.

If the patternmaker is not familiar with metalworking, then it may be wise to combine the facilities of toolmaking and pattern rooms and allow metalworking and woodworking specialists to pool their knowledge. Companies that lack the proper facilities or personnel for wood tooling may find it to their economic benefit to subcontract all wooden jigs and fixtures to outside pattern or woodworking shops.

Wood Selection

The hardwoods maple and birch are usually the best for heavy-duty tooling construction, for these two woods possess many favorable qualities. Any of the manufactured woods will also be a successful choice. But for jigs and fixtures that are subjected to light duty, it is possible to use pine, a softwood that is light in weight, straight-grained, and easily worked and which will not warp, shrink, or check as readily as some other woods.

In case of uncertainty about the most desirable wood to use for a given tooling problem, refer to Table IX–1, Properties of Wood.

Joining Methods

Jigs and fixtures made of wood will almost always include one or more joints in their construction, and their details may be joined with glue or mechanical methods such as nails, screws, or plates. For increased strength in a joint, a combination of glue and one of the mechanical methods should be used. Although there are many types of joints that can be used for wood construction, some of them are simply variations or combinations of a few basic types — butt, rabbet, dado, mortise-and-tenon, miter, and dovetail. Each of these joints has a particular purpose and can be applied, adapted, or reinforced to suit the varying requirements of different tools.

Butt. The butt joint is the simplest and most popular of the woodworking joints. Unfortunately, it is the weakest joint when it is not reinforced. Figs. 9–7a and b illustrate the end-to-end and end-to-side butt joints, respectively. The end-to-end joint is difficult to glue and rarely meets requirements of strength

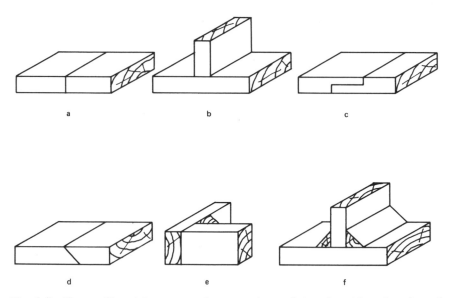

Fig. 9–7. Types of butt joint construction—a, end-to-end; b, end-to-side; c, lap; d, scarf; e and f, reinforced end-to-side.

and durability because the glue is rapidly absorbed by the end grain. The lap (c) and scarf (d) joints are methods of joining the stock so that end-grain absorption is more or less avoided and more area of contact is provided for the glue. The end-to-side joint should be reinforced as shown in Figs. 9–7e and f for maximum strength.

Rabbet and Dado. The rabbet and dado joints are very similar to each other in construction. For a rabbet joint, the wood pieces are joined at the ends as shown in Fig. 9–8a. The dado joint consists of a groove made across the grain of a piece of wood into which another member is fitted as shown in b. The dado joint can be made even more rigid by cutting additional shoulders as shown in Fig. 9–8c. The combination of dado and rabbet (d) forms a highly satisfactory corner construction.

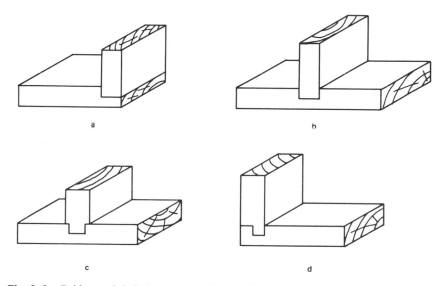

Fig. 9–8. Rabbet and dado joints — a, rabbet; b, dado; c, dado reinforced with shoulders; d, combination dado and rabbet.

Mortise-and-Tenon. When great strength and rigidity are required, the mortise-and-tenon joint shown in Fig. 9–9a is used. The joint may be cut into the two members as shown in a, or dowel pins may be used as shown in b. The latter form is often referred to as a dowel joint. The mortise-and-tenon joint is difficult to make accurately and should be used only when a simpler method will not serve.

Miter. A miter joint should be used when the end grain must not be exposed. Fig. 9–10 illustrates its construction. If it is not possible to use a corner reinforcement (a), the joint may be strengthened by the use of wood dowels as shown in b or corrugated metal fasteners as shown in c.

Dovetail. The dovetail joint shown in Fig. 9–11 is rarely used in wooden jig and fixture construction because it is the most difficult joint to make. However, when it is properly fitted and glued it is probably the soundest of all the basic joints.

Mechanical Reinforcements. Mechanical reinforcements should be used for tools that may be subjected to hard use, because it is estimated that reinforced joints are 50 percent stronger. The various reinforcements shown in Fig. 9–12 are: a, mending plates; b, flat corner irons; c and d, outside and inside corner braces; and e, T plates.

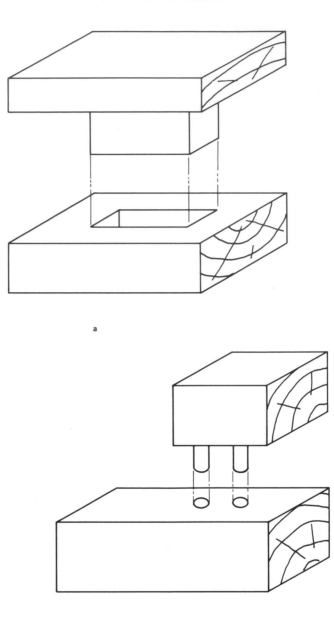

Fig. 9-9. Mortise-and-tenon joints — *a*, cut; *b*, doweled.

Hardware for Wood Tools

A jig or fixture will seldom be totally constructed of wood. Usually, metal components such as locators, stops, wear plates, and conventional hardware must be used to maintain part orientation and reduce wear. Many of these metal

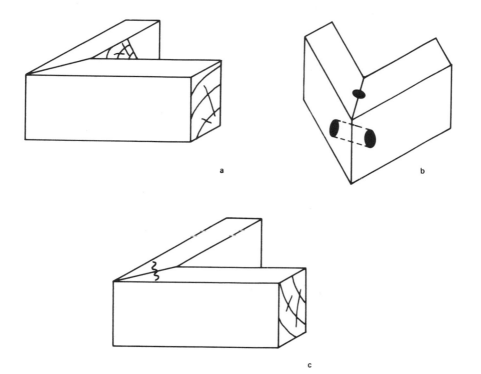

Fig. 9–10. Reinforced miter joints— a, with corner block; b, doweled; c, with corrugated metal fastener.

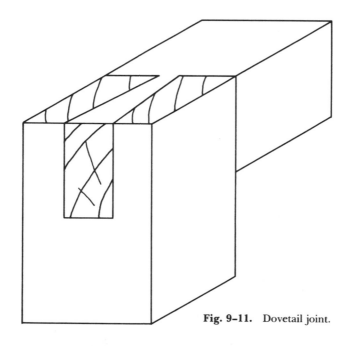

Fig. 9–11. Dovetail joint.

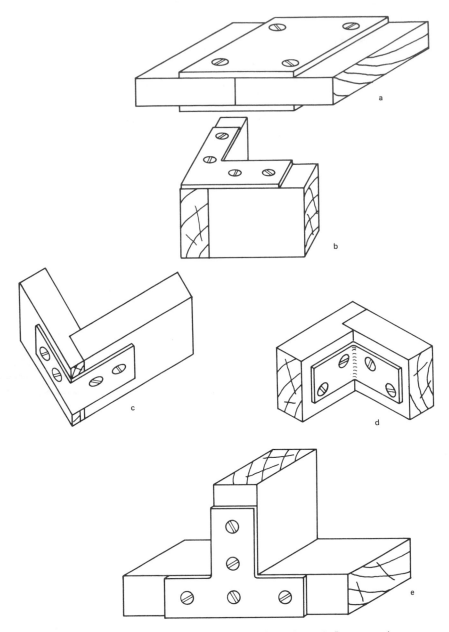

Fig. 9–12. Mechanically reinforced joints — *a*, mending plates; *b*, flat corner iron; *c*, outside corner brace; *d*, inside corner brace; *e*, T plate.

components are available commercially, and they should be used whenever economically feasible. Generally, most standard commercial hardware for conventional tooling can be adapted to wood tools, but in recent years, hardware manufacturers have felt the need for components specifically designed for wood construction and have begun to produce parts for that purpose. Examples of such parts are the drill or guide bushings shown in Fig. 9–13. The bushing on the left has straight serrations on its outer circumference so that it can be press fit securely into the wood. The bushing on the right is inserted into the wood and embedded with a filler material. With its opposed serrations and center groove, the latter ensures positive anchorage in the wood. Both bushings are commercially available at American Standards Association (ASA) standards.

Fig. 9–13. Standard commerical drill bushings for wood tooling. (*Courtesy, Ex-Cell-O Corporation*)

Because tooling hardware for limited-production tools does not need to be hardened, as does that for mass-production tools, stops and locators for wood tools in limited production may consist merely of standard screws with nuts and washers as shown in Fig. 9–14. However, the ends or locating surfaces of the

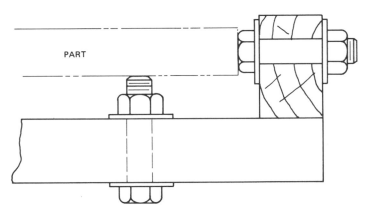

PART

Fig. 9–14. Standard screws used as stops and locators for wood tooling.

screws should be machined or ground after tightening in order to align them properly. Washers are necessary to prevent the screws and nuts from depressing the wood surface. If a larger locating area is desired, the screw and nut assembly can be replaced by a sheet metal plate bonded to the wood or fastened to it with flat-head wood screws.

Steel threaded inserts must be embedded into the wood if components such as studs and machine screws are to be attached. Inserts can be specially made, or standard commercial Allen-type socket nuts like that shown in Fig. 9–15 can be used. These nuts have been used in wood tool construction with outstanding results. They are knurled on their outer circumference, and when press fitted they lock into the wood to prevent rotation. For added strength, glue can be applied to the outside or knurled portion of the nut prior to assembly.

Fig. 9–15. Commercial socket nut for use with wood tooling. (*Courtesy, Allen Manufacturing Company*)

Wood tooling may occasionally require the use of dowel pins for alignment and other purposes. The standard cylindrical dowel pins frequently used for conventional tooling will not work in wood, however. Grooved or spring pins must be used instead.

A wood drill jig incorporating the various tooling hardware is shown in Fig. 9–16. The jig consists of a plywood base (A), a top bushing plate (B), and two wood side spacer rails (C). A sheet metal plate (D) is bonded to the top of the wood base to reduce wear and depression as the part or work is clamped. The base, top, and side rails are joined with wood screws and the four screw assemblies which also serve as jig feet. A spring pin (E) is used as an end stop. Side locators (F) also consist of screw assemblies. A hand knob clamp (G) and thumb screw (H) are standard tooling components used with socket nuts (I) as threaded inserts. Drill bushings (J) are the standard type used for wood tooling.

Wood Protection

Protecting a wooden jig or fixture by coating it to seal its open cells or pores is an important step in its completion. The sealing will reduce swelling and warpage caused by the absorption of atmospheric moisture or the oils and coolants used in manufacturing. For normal conditions, two to four coats of boiled linseed oil will provide adequate protection. Be extremely careful to

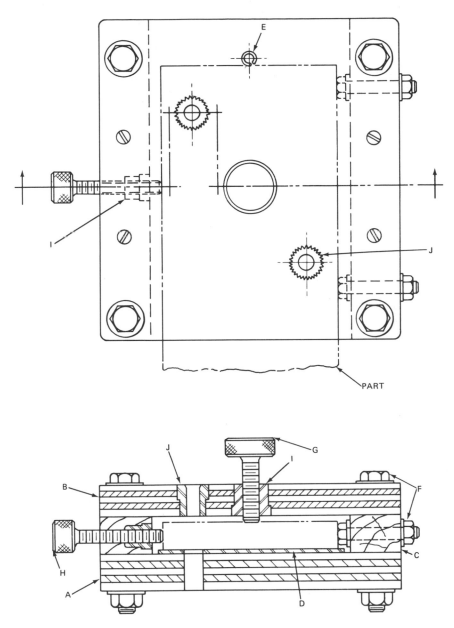

Fig. 9–16. Wooden drill jig incorporating standard tooling hardware.

apply the linseed oil evenly with a clean cloth. Then rub the wood hard and vigorously with a dry, clean cloth. The oil should never be applied with a brush or by dipping. Under more rigorous conditions, the wood should be sealed with shellac, lacquer, or a high-grade wood sealer for maximum protection.

ECONOMICS OF WOOD TOOLS

Wood used for tooling can provide savings of as much as 75 percent of the cost of equivalent steel tools. This savings does not include a savings in handling costs because wood tools are approximately one-sixth the weight of their conventional equivalents. Since low-cost handling equipment can be used in both the tool fabrication and the production departments, savings are further compounded.

As an illustration of these points, consider the plate drill jig shown in Fig. 9–17. The jig was designed to locate six .875-in holes in a large machine casting.

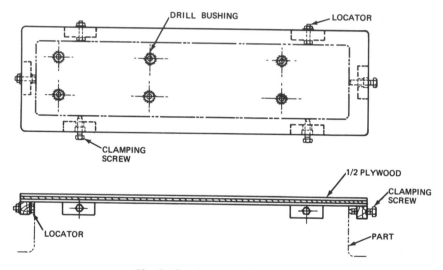

Fig. 9–17.   Large wooden plate jig.

Because of its size (18 × 48 in), the jig would have been too heavy for one man to place over the casting if it had been made of steel. Other lightweight materials were too expensive and could not be obtained in time to meet the production schedule, So wood was chosen for the construction material. A sheet of .5-in exterior plywood was used as the main structure of the jig, the bushing plate, and the entire jig required only a few hours to build at a fraction of the cost that would have been required for an all-steel jig.

Savings on wood tools are usually greatest on large structural tools for which material represents a major proportion of tool cost. Generally, tooling for small parts should be examined carefully, for it may be found that such tooling will not justify wood construction. Industries that produce small items do not normally profit from wood tools unless their product is complicated or has complex contours, in which case the easy formability of wood may provide savings.

The milling fixture shown in Fig. 9–18 is an excellent example of a relatively small tool that was economically fabricated from wood. The fixture was

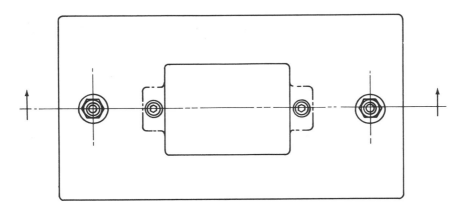

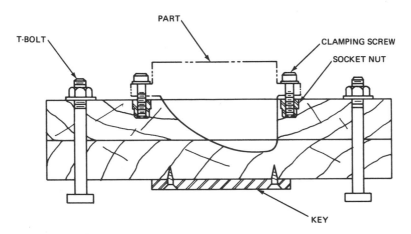

Fig. 9–18. Wooden milling fixture designed to hold an irregular casting.

designed to hold an irregular casting, and a cavity or nest with the contour of
the casting was cut into the base of the fixture as shown. The cavity was formed
in the wood easily, but the same cavity formed in a steel fixture would have
required expensive equipment and would have resulted in a much higher tool
cost. The wood fixture was built for $45. The low bid for an equivalent steel
fixture was $103. The decision to use wood in this case, although the part to be
machined was small, resulted in a savings of $58.

If a complete wood fixture is impractical in a similar case, then a wood insert with the configuration of the casting may be incorporated into an otherwise all-steel fixture and thereby reduce costs.

The bracket illustrated in Fig. 9–19 is another item that was considered ideal for wood tooling because of its length, the limited quantity of 250 pieces ordered, and the short lead time—one week. The bracket was received as a

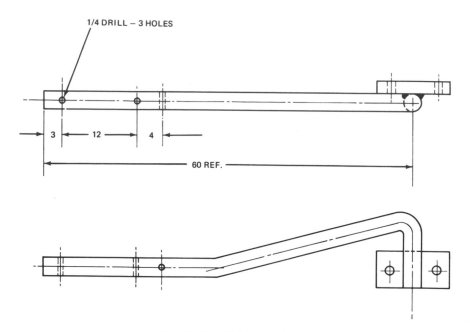

Fig. 9–19.   Welded bracket.

welded construction with the two holes in the end plate. To complete the bracket, three holes had to be drilled into it, and a wooden tumble jig, shown in Fig. 9–20, was designed and fabricated to do the job. The jig consisted of a pine base (A), a maple front locating block (B), and an attached bushing plate (C). The bushing plate contained two wood-type drill bushings and a standard hexagon screw (D) for a clamp. The locating block also incorporated a bushing for the drilling of the third hole, for which the jig had to be turned on its side. The maple back block (E) was required to align the end plate. Two standard screws (F) were used for clamping the long end of the bracket, sheet metal plates (G) were used as jig feet, and Allen nuts were used as threaded inserts for the screws.

Because of the short lead time, the tumble jig was built from a preliminary sketch in one day at a total cost of $63. An equivalent steel jig would have required more fabrication time and considerably more expense. In addition, it would have been difficult to tumble an all-steel jig of such a size.

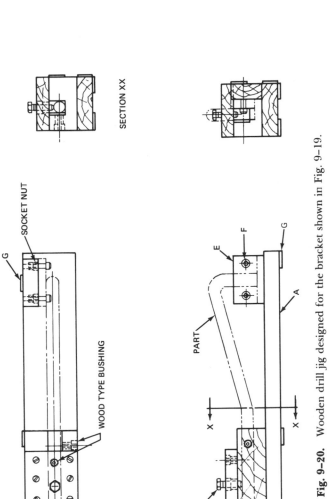

SECTION XX

SOCKET NUT

WOOD TYPE BUSHING

PART

Fig. 9–20. Wooden drill jig designed for the bracket shown in Fig. 9–19.

chapter

10

epoxy plastic
tool construction

Epoxy plastic tooling possesses several advantages. Jigs and fixtures made of plastic are being used in increasing numbers, particularly in limited production. The intelligent development and use of plastics for tool fabrication can provide substantial savings in both time and money.

Plastic tooling is no panacea; it will not solve every tooling problem nor will it provide savings where there are none to be had. As with other tooling, plastic tools must be carefully designed, practical to build, functional in operation, and dependable in service. Plastic has certain limitations as a tool fabrication material, and these should be clearly understood.

This chapter seeks to aid the tool engineer in evaluating tooling plastics for his needs. First, tooling plastics are defined and described, and then their advantages and limitations are examined. Next we take a look at the design considerations involved in plastics applications, and the chapter closes with a discussion of the methods of shaping plastic and attaching hardware to assemble the tools.

PLASTICS

The adjective *plastic* refers to any material capable of being molded or shaped. A plastic, as we shall use the term, refers to a group of organic, synthetic materials which possess the characteristic of plasticity to a greater or lesser degree. Although not all plastics used in manufacturing are easily formed or molded, the term *plastic* is particularly suitable for the plastic materials used for tooling, since good formability is one of their most important advantages.

Plastic resins do not occur naturally, but are produced by chemical reaction. The physical nature and properties of plastics depend on their molecular structure and the combinations of heat, pressure, and chemical catalysts used to form and cure them. Plastics are classified in two groups according to their reaction to heat. Although there are some resins that exhibit characteristics common to both groups, the majority of plastics are known as either (1) thermoplastic, or (2) thermosetting. Thermoplastic materials can be melted by heat, formed, and allowed to cure by cooling. After cooling they can be remelted and re-formed. Thermosetting plastics, on the other hand, may be softened by heat during forming but cure to their solid form with continued heating. Thermosetting resins cannot be remelted after they have cured, and the continued application of heat will destroy them.

113

Epoxy Resins

Epoxy resins are the principal type of plastic used for tooling construction. Phenolics and polyesters are sometimes used, but none of them have the unique combination of properties that makes epoxy resins suitable. Basic epoxies are thermoplastic and have no value until they are cured by the addition of an activator called a catalyst, converter, or hardening agent. Activators transform the resin into thermosetting material of high chemical and thermal resistivity, high adhesive ability, high strength, easy machinability, low shrinkage, and stability under working conditions. For the actual physical properties of cured epoxy resins, see Table X–1 (the addition of most fillers to the resin considerably increases the values).

Table X–1. General Physical Properties of Epoxy Resin (Pure State).

Tensile Strength	10,000–12,000 psi
Compressive Strength	18,000–22,000 psi
Flexural Strength	18,000–20,000 psi
Impact Strength (ft-lb/in notch)	.5–1.5
Rockwell Hardness$_M$	100–125
Cure Shrinkage (in/in)	.0002–.001
Heat Distortion Point	180° F

Epoxy resins are available from formulators who purchase the resins, catalysts, hardeners, and fillers in large quantities and prepare the materials for specific uses. They modify the resins, add various fillers such as aluminum, steel, stainless steel, brass, bronze, and lead, and market them in convenient packaged units with specific instructions for their use. It is highly important that the instructions given by the formulators regarding mixing, preparation, and curing time are followed carefully.

Fillers

Fillers are used both to reduce tool cost by extending the epoxy and to enhance the properties of the plastic. Formulators supply epoxy resin without fillers, and the user may supply and mix his own. The filler most often used in laminated epoxy is fiber-glass cloth or cotton cloth. For cast epoxy, aluminum grains, spheres, and needles are the most common fillers. Aluminum grain filler gives excellent dimensional stability to the cast plastic, and the mixture can be machined without difficulty. Iron, iron oxide, steel, silicon carbide, and other materials are also used as fillers for cast epoxy.

For large tooling, inexpensive fillers such as sand and gravel can be used. They give the casting good dimensional stability, but they make machining difficult. Ground walnut shells can be used as a filler when the quality of high heat dissipation is not required. Crushed glass, limestone, ceramic, quartz, and granite are also inexpensive fillers. The addition of hard fillers such as carbides or glass will increase the abrasion resistance of the surface of the tool.

The ratio of filler to epoxy depends largely on the application, but a pourable consistency must be maintained in order to produce a solid mass without air voids when the tool is formed. The advantages and limitations of various fillers used in the fabrication of tooling are shown in Table X-2. This table should be used only as a guide or for general information; special applications or conditions may alter the ratings.

Table X-2. Properties of Plastic Tooling Filler

Filler	Method	Advantages	Disadvantages
Fiber Glass Cloth	Laminating	High strength Low shrinkage Dimensional stability	Slow process High labor cost
Fiber Glass Woven Rovings	Laminating	Requires fewer laminates High strength	Difficult to wet out Slow process
Fiber Glass Mat	Laminating	High strength Low shrinkage	Low glass content
Milled Glass Fibers	Paste and casting	Increased strength	
Cotton Floc	Paste	Smooth texture Easy to mix Easy to machine	Low strength High shrinkage
Metal Spheres or Grains	Casting	Low cost High strength Reduced exotherm Easy to mix and pour	
Walnut Shells (crushed)	Casting	Low cost Easy to machine	Unstable High moisture content High shrinkage
Lava, Slag, Coke	Casting	Low cost Reduced exotherm	Low strength Difficult to machine Dirty
Sand and Gravel	Casting	Low cost Reduced exotherm High strength Dimensional stability	Difficult to machine

(Courtesy, Ren Plastics)

ADVANTAGES AND LIMITATIONS OF PLASTICS

The application of epoxy plastics to jigs and fixtures requires the same approach that would be given any other material. The properties of plastics listed in this section should be studied carefully and their limitations should be analyzed in relation to the tooling application.

Lower Cost

The major advantage of plastic tools for limited production is that capital tooling investments for them are often 50 percent less than investments for conventional tools. This reduction in tooling costs can often allow the manufacture of products that are complex or oddly shaped and otherwise could not be produced economically.

Shorter Lead Time

Fabrication time for plastic jigs or fixtures is considerably reduced because they can be easily formed in the desired shape without machining. When tooling can be fabricated quickly, a product can be put into production with a minimum of lead time.

Plastic tooling can even be used for high-volume products when lead time is limited. The volume of production may warrant conventional tooling, but epoxy tools may be used until they can be replaced by the permanent tools. This dual-tooling method is expensive, but it can shorten the time required to get production started.

The reduction in lead time resulting from the use of epoxy tooling is difficult to estimate, but one automotive manufacturer claims lead-time savings of as much as 70 percent through the use of such tools.

Light Weight

One cubic foot of cured plastic will weigh from 70 to 90 lbs. In comparison, the same volume of steel will weigh nearly 500 lbs and aluminum, 160 lbs.

The light weight of plastic tooling can provide substantial benefits in production operations when jigs and fixtures require considerable handling. Heavy metal tooling may warrant replacement with epoxy tooling even though the latter may be more expensive.

Formability

Epoxy plastic can be easily formed to complex configurations. The form is usually laminated or cast around a master pattern which may be a prototype part, a production part, or a scale model made of wood, plaster, plastic, or metal. Either lamination or casting will reproduce the contours of the master pattern without complicated and expensive machining. On the other hand, reproduction of a complex configuration in steel would require expensive equipment and high costs.

Corrosion Resistance

Epoxy plastic materials remain unaffected by most of the corrosive gases and liquids used in manufacturing. They are also impervious to grease and oil and are not affected by bad weather when they are stored outdoors.

Ease of Modification

Because of the ease with which cured epoxies can be modified, industries that experience continuous product changes should consider plastic tooling. Alterations can be made without completely scrapping the tool, for epoxy material can be added, removed, or relocated to suit the situation. During the design and development stages, it is relatively simple to modify the design of a tool to correct any errors that may appear during functional testing.

Tools made of plastic can also be easily repaired with a minimum of cost and delay. Curing time for repairs or modifications can be shortened by directing a heat lamp or other heat source on the repaired spot.

Mechanical Properties

When the most in mechanical capabilities is required, plastic tooling will probably be of limited use. Although epoxies are lightweight, corrosion resistive, and easily formed, they do not have the strength, durability, and dimensional stability of other materials.

Their mechanical properties are even further reduced at high temperatures. To date, the highest temperature at which any tooling epoxies can maintain dimensional stability and strength under pressure for a limited time is 500° F. For all practical purposes, plastic tools should not be used when temperatures exceed 400° F.

Accuracy

Since there is seldom any finishing required on the face of plastic tools, the accuracy cast into the tool will be the accuracy imparted to the product. The form, mold, production part, or fixture on which the tool is laminated or cast must possess the same degree of accuracy as that required of the tool.

The accuracy that can be expected of epoxy jigs and fixtures is approximately .002 in/ft, and drill bushings can be located within ±.0004 in. These close tolerances are only possible when the tool is laminated.

Toxicity

Many of the tooling epoxy plastics used today present health problems. Workers may develop severe dermatitis or respiratory diseases after prolonged exposure to epoxy catalysts unless proper precautions are taken. Although the packaging developed by resin suppliers and formulators has to some extent minimized this problem, the following procedures in safe handling should always be followed:

1) Storage rooms and workrooms containing epoxies should be well ventilated and clean. Exhaust fans should be located at the same level as the work and should draw epoxy fumes away from the worker and not up into his face.
2) Gloves should always be worn by persons using epoxies.
3) Protective clothing should always be worn, and it should be changed frequently.
4) Workers who prepare and form epoxies should wash frequently with mild soap and water.

DESIGNING EPOXY TOOLS

Epoxy tools are not complex or difficult to design although their design differs considerably from conventional tooling design practice. Epoxy jig and fixture design is limited only by the ingenuity of the designer and the physical properties of the material.

The first step in epoxy jig and fixture design is to decide on the most suitable type of construction by evaluating the advantages and limitations of the various construction methods listed in Table X–3. The properties of cured epoxy plastic depend to a great extent on these construction methods. Since every tool meets different service conditions, the method of fabrication used must meet those specific requirements.

Table X–3. Plastic Tooling Construction Methods.

	Dimensional Stability	Shrinkage (during cure)	Low Weight	Labor Cost
Laminate	Excellent	Excellent	Excellent	Poor
Surface Cast (Metal Core)	Fair	Fair	Poor	Fair
Mass Cast	Poor	Poor	Fair	Excellent
Paste	Satisfactory	Satisfactory	Satisfactory	Satisfactory

	Material Cost (per lb)	Strength	Toughness
Laminate	Poor	Excellent	Excellent
Surface Cast (Metal Core)	Fair	Satisfactory	Satisfactory
Mass Cast	Excellent	Fair	Poor
Paste	Satisfactory	Poor	Fair

(Courtesy, Ren Plastics)

Selection of the epoxy material to be used is the next step. The proper material is very important and should be selected carefully. There are many grades of epoxies available to suit practically any tooling purpose, and formulators usually provide a material selection chart of their various grades. For special complex cases, an epoxy formulator's representative should be consulted.

Epoxy jigs and fixtures can usually be fabricated without drawings. In other cases, freehand sketches or verbal instructions may be sufficient. In all cases, however, a permanent record of the source and grade of the epoxy used should be kept in case the tool requires modification or repair. For complex tooling, formal drawings are often necessary. Fig. 10–1 shows a typical drawing of an epoxy drill jig.

Because the coefficients of thermal expansion of plastic and metal differ, the epoxy cast tool should be designed to avoid the bonding or mechanical fastening

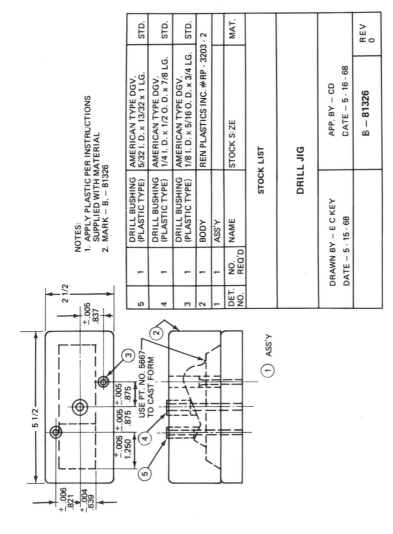

Fig. 10-1. Drawing of an epoxy drill jig.

of long lengths of metal. Also, metal weldments should always be normalized before they are attached to a tool.

FABRICATING EPOXY TOOLS

Basically, construction with epoxy plastics can be divided into four general categories:
1) Lamination
2) Surface casting
3) Mass casting
4) Paste construction.

All these methods may be used singly or in any combination. To determine which method would be best for a particular application, seven factors should be considered for each construction category. The seven factors are:
1) Dimensional stability
2) Shrinkage
3) Weight
4) Labor Cost
5) Material cost
6) Strength
7) Durability.

Table X–3 lists all seven selection factors with respect to the four construction categories. The table should be used only as a general guide, however, for conditions of a particular application may alter the factors.

A pattern or mold must be built for any of the four construction methods. The pattern is usually made of wood, although plaster, metal, or plastic can be used. In some cases, a prototype or production part can be used if such is available.

All patterns must be properly prepared before they can be used. A wood pattern must be finished with a lacquer sealer and prepared by the following steps:
1) Clean the surface of the pattern thoroughly with acetone, alcohol, or another commercial solvent.
2) Apply a high-grade, carnauba-based paste wax and rub it into the pattern thoroughly with a lint-free cloth.
3) Apply a coat of parting agent by spray or brush.
4) After the parting agent has dried, apply paste wax and wipe it off with a lint-free cloth. Do not rub the wax in.

Plaster patterns are sealed with lacquer and allowed to dry, after which steps 2, 3, and 4 above are followed as for wood patterns. Any common plaster is suitable for plaster patterns.

Plastic patterns are treated in the same way except that all foreign matter should be removed from the surface before treatment, and lacquer sealer is not necessary.

For metal patterns with machined finishes, all surface grease should be removed with a suitable solvent. After the pattern has been cleaned, three coats of carnauba paste wax are applied. Each coat should be rubbed with a lint-free cloth before the next coat is applied.

Metal patterns which have not been machined must also be cleaned with solvent. Then carnauba paste wax is applied and the excess is removed with a lint-free cloth. Next, one brush coat or three light spray coats of a parting agent are applied. After the parting agent has thoroughly dried, a coat of the wax should be applied and the excess removed.

After the pattern has been prepared, the lamination, casting, or molding of the tool may begin.

Laminating

In the laminating method of construction, alternate layers of glass cloth and epoxy are built up on a form until the desired thickness is obtained. Fig. 10-2 shows how the cross-section of a laminate looks.

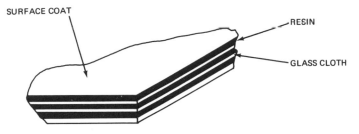

Fig. 10-2. Laminated construction cross-section.

After the pattern has been prepared for a laminated construction, a surface coat or gel is applied. The purpose of this coating is to provide enough material to prevent the first layer of glass cloth from touching the face of the tool, and the type of gel to be used will depend on the tool being fabricated. Special gels are available in a variety of forms with varying degrees of hardness, resilience, and abrasion resistance, but common laminating resin can be used. The first coat should be applied with a brush or trowel and allowed to cure to a tacky state before lamination is begun. A glass paste mixture made of laminating resin mixed with chopped glass fibers and flock should be used to fill sharp corners and depressions in the pattern which may cause voids or bubbles under the first layer of glass cloth.

The glass cloth is then laid over the tacky resin and stippled into place with the end of a brush. Alternate layers of laminating resin and glass cloth are applied until the desired thickness is achieved. No more than eighteen layers should be applied at any one time, for the heat generated when an excessive amount of plastic cures may cause the finished tool to shrink or warp. If excessive heat is noted at any time, work should be stopped until the material has cooled.

All voids and air bubbles must be excluded from between layers of laminates, and the glass cloth must be fully impregnated with resin. If a hardened surface is glossy it should be sanded before the next resin layer is applied. The success of the final tool depends to a large extent on the care with which these steps are performed.

If a framework for increased strength or rigidity is to be added, it should be included immediately after lamination is completed. The framework can be made from any material such as glass-laminated honeycomb structures, plastic tubing, paper-based phenolics, sheet stock, plywood, or metal. For best results, the framework should be of the same material as the tool facing, because age, weight, and changes in temperature and humidity can all affect a plastic tool with dissimilar framework and face. Any expansion within the framework can cause the laminated tool facing to warp. After the lamination is completed, the liquid epoxy and the glass cloth solidify into a strong, rigid form. A drill jig constructed by the lamination method is shown in Fig. 10-3.

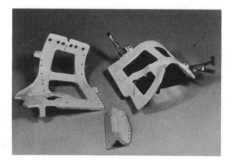

Fig. 10-3. Laminated plastic drill jig and scribe template. (*Courtesy, Ren Plastics, Inc.*)

Jigs and fixtures of laminated construction are hard to surpass when accuracy, low cost, low fabrication time, strength, and light weight are major considerations. Laminated tools possess the best dimensional stability of any of the plastic tooling types, and for this reason they are widely used in industry as gages and checking fixtures.

Surface Casting

Surface-cast tooling is constructed by casting an epoxy face approximately .125 to .75 in. thick on a prefabricated core. The thickness depends on the size of the tool and the method of surface casting. The core can be made of any suitable material.

Surface casting is widely employed for sheet metal forming dies and for numerous jig and fixture applications. The dimensional stability of surface-cast tools is somewhat less than that of laminated tools, but it is sufficient for most applications. A typical surface-cast form is shown in Fig. 10-4.

There are three methods of surface casting—pour, squash, and pressure. The core for any of these methods, but especially for the squash method, should always be tried in the mold before the epoxy is added. This trial is to check for points at which air could be trapped by the core and push epoxy away from the core's face. If air might be trapped, vent holes can be drilled through the core, or the mold might be elevated slightly on one side to release the air. The core should always be placed above the form so air bubbles will not be formed on the working surface of the mold. Epoxy should be poured into the mold from one position and in one steady stream to prevent voids from being formed in the tool.

Fig. 10-4. Typical surface-cast epoxy plastic form. (*Courtesy, Ren Plastics, Inc.*)

Pour Method. In the pour method the core is suspended with ample clearance between it and the working face of the pattern. The edges of the space between the mold and the core are sealed, and epoxy is then poured through spouts or an open end of the mold into the space. This method is illustrated in Fig. 10-5.

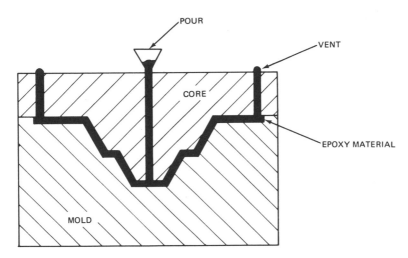

Fig. 10-5. Surface casting, pour method.

Squash Method. In the squash method a core is placed into the cavity of a pattern mold partially full of liquid or paste epoxy and is allowed to settle to a predetermined position as shown in Fig. 10-6.

Pressure Method. As in the pour method, the core for the pressure method is suspended over the mold and the edges between the mold and the core are sealed. Liquid epoxy is then forced into the mold at a low point and is allowed to vent at a high point as shown in Fig. 10-7. This method requires a pressure pot or other means of forcing the epoxy into the cavity. The pressure pot should be located close to the mold to eliminate an excessive amount of hose line,

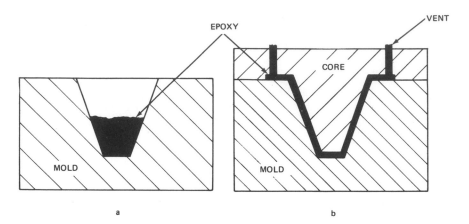

Fig. 10–6. Surface casting, squash method — *a*, mold partially filled with liquid epoxy; *b*, core lowered into mold.

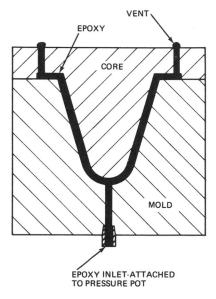

Fig. 10–7. Surface casting, pressure-pot method.

and the pressure pot and system must be thoroughly cleaned immediately after the epoxy has been pumped into the mold.

Mass Casting

Mass casting is the construction method in which epoxy liquid is poured into a prepared mold as shown in Fig. 10–8. No core is used. The liquid plastic must be poured slowly to avoid the development of air pockets, and when castings over six inches thick are poured, one of the fillers listed previously should be added to the epoxy to add strength and reduce shrinkage in the casting. The casting with filler will also require less epoxy and allow material costs to be reduced. The mass-cast vacuum fixture shown in Fig. 10–9 is typical of the type of tool constructed by this method.

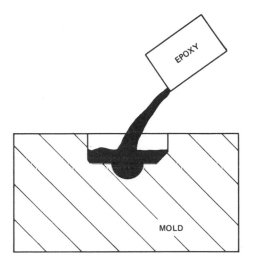

Fig. 10–8. Mass casting.

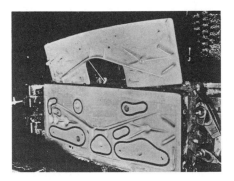

Fig. 10–9. Epoxy mass-cast vacuum fixture. (*Courtesy, Ren Plastics, Inc.*)

Although the simplicity of the mass-casting method implies the greatest savings in time and material of all the methods, this is often not the case. Mass casting does provide some savings, for it eliminates the cost of the core needed for surface casting. But since the preparation of a model or mold is the same in mass casting as in the other casting methods, the only other savings provided are due to lower labor cost. Unless savings are definitely possible, mass casting should not be used, for it has unfavorable characteristics that reduce its value.

The greatest limitation of mass-cast tools is their low strength. High shrinkage in heavy castings because of the heat generated in curing is another unfavorable characteristic of this method. Recent material developments have improved these properties to some extent, however, and future discoveries in this field may eliminate them and make epoxy mass casting a highly efficient toolmaking method.

Paste Construction

Paste constuction is the method in which an epoxy paste made of epoxy resin, filler, and hardening agent is applied to the tool pattern with a spatula, putty knife, or caulking gun. Fig. 10–10 shows the procedure for preparing and applying epoxy paste.

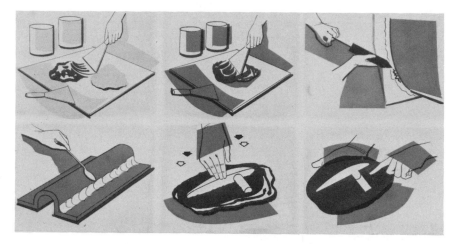

Fig. 10–10. Steps in the preparation and application of epoxy paste. From the top, l. to r.: (1) Measure proper proportions of resin and hardener into two piles on flat surface. Note use of two putty knives. (2) Mix resin and hardener thoroughly. (3) Use a caulking gun to apply the mixed paste, or (4) apply the mixed paste as a fillet material to fill sharp radii. (5) Use paste material in a squash method to make impression of any definite form. (6) Final impression with form removed after paste has set. *(Courtesy, Ren Plastics, Inc.)*

Because epoxy paste is nonflowing, it does not have to be formed on a level surface as liquid epoxy does. It can be applied to vertical surfaces, and it generally does not have to be held in position with special equipment. Epoxy paste is also perfectly suited to the squash method of surface casting.

Epoxy paste has wide use in almost every phase of the metalworking industry, but especially in tooling construction. One of its uses is in the construction of fixtures that locate or hold parts of complex or irregular contour. Fig. 10–11 shows a lathe chuck with epoxy paste jaws used to hold an irregular casting during machining. To construct the jaws, a casting of the part was coated with a parting agent, and epoxy was formed like modeling clay around half of the part and allowed to cure. The same procedure was then followed to form the other half-section. The two sections were then mounted in a two-jaw chuck. The duplication of these jaws in metal would have been difficult and costly.

Another favorable quality of epoxy paste is that it can be worked with

Fig. 10–11. Epoxy lathe jaws for holding an irregularly shaped part. (*Courtesy, Devcon Corp.*)

limited equipment. Although its dimensional stability is slightly less than that of laminated tools, that stability is well within normal tolerance requirements.

Epoxy paste is commercially available with a variety of fillers for special applications. Among the common fillers are aluminum, iron, and stainless steel. Chopped glass fibers, cotton floc, asbestos, and aluminum spheres can also be added as fillers.

Hardware Attachment

Metal components such as bushings, plates, clamping studs, wear pads, etc., are required for epoxy tooling just as for any other type of tooling construction material. In cast epoxy tooling, the metal components are easily cast into the tool face. In laminated tooling a hole slightly larger than the component must be drilled and epoxy paste must be used to hold the part in the cavity. To provide a strong bond, the metal component must be shaped to form an interlock with the plastic. Fig. 10–12 illustrates some methods of interlocking metal parts in epoxy. Since most epoxies can be easily drilled and tapped, metal components can also be mechanically fastened with screws.

Conventional drill bushings should never be used in epoxy jigs. Bushings specially designed for plastic tooling are available and should be used. Note a few bushings of this type shown in Fig. 10–13.

The method of inserting a bushing in plastic tooling differs from the normal method. When a jig is fabricated by one of the casting methods, the drill bushing must be positioned in the exact location before the liquid epoxy is poured. This positioning can be done with a sample part. The bushing hole should be aligned with the corresponding hole in the part by a locating pin the exact size of the hole as shown in Fig. 10–14. Normal epoxy casting procedures are then followed.

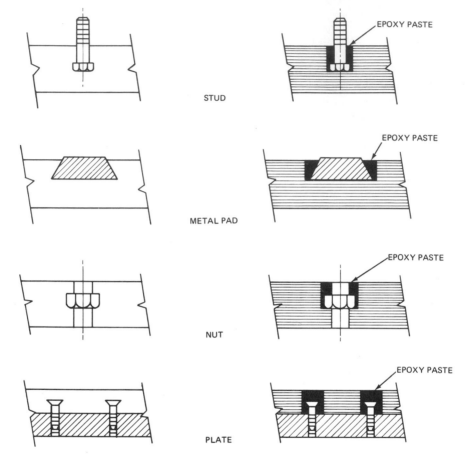

Fig. 10–12. Methods of attaching metal parts to epoxy tooling.

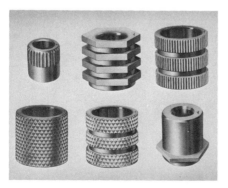

Fig. 10–13. Commercial drill bushings for epoxy tooling. (*Courtesy, American Drill Bushing Company*)

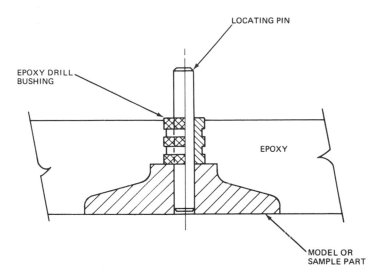

Fig. 10–14. Method of locating and bonding drill bushings in cast epoxy.

For laminated jigs, a model or sample part is prepared and normal laminating procedure is completed. After the tool has cured and before it has been removed from the model, pilot holes for the drill bushings are drilled with the model as a jig for their location. The tool is removed from the model, and the holes are then redrilled to a diameter .0625 in. greater than the outer diameter of the bushings. Finally, the laminated form is replaced on the model, locating pins are inserted through the oversized holes into the holes in the model, and bushings are placed over the pins and into the oversized holes. Epoxy paste is then cast into the holes around the bushings as shown in Fig. 10–15.

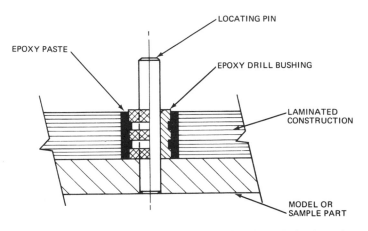

Fig. 10–15. Method of locating and bonding drill bushings in laminated epoxy.

additional readings

The following list of books, articles, papers, and other materials is recommended to both the novice and the experienced tool designer or toolmaker for sources of information on fundamental principles of jig and fixture design as well as additional sources on materials, design, and fabrication of limited-production jigs and fixtures. The list is not meant to include all the literature on these subjects.

BOOKS

American Society of Tool and Manufacturing Engineers. *Fundamentals of Tool Design.* Englewood Cliffs, N.J.: Prentice-Hall, Inc., 1962.
———. *Handbook of Fixture Design.* New York: McGraw-Hill Book Co., Inc., 1962.
———. *Tool Engineers Handbook.* New York: McGraw-Hill Book Co., Inc., 1959.
Cole, C. B. *Tool Design.* Chicago: American Technical Society.
Donaldson, Cyril, and George H. LeCain. *Tool Design.* New York: McGraw-Hill Book Co., Inc., 1957.
Gross, W. H. *The Story of Magnesium.* Cleveland: American Society for Metals, 1949.
Howe, Raymond E., ed. *Producibility/Machinability of Space-Age and Conventional Materials.* Dearborn, Mich.: American Society of Tool and Manufacturing Engineers, 1968.
Jones, Franklin Day. *Jig and Fixture Design.* New York: Industrial Press, 1955.
Kempster, M. H. A. *Principles of Jig and Tool Design.* New York: Hart Publishing Co., 1969.
Koch, Peter. *Wood Machining Processes.* New York: The Ronald Press Company, 1964.
Lascoe, O. D., ed. *Proceedings of the Seventh Annual Seminar on Plastics for Tooling, June 10–11, 1965.* Purdue University School of Industrial Engineering.
Riley, Malcolm W. *Plastics Tooling.* New York: Reinhold Publishing Corp., 1961.
Roberts, C. Sheldon. *Magnesium and Its Alloys.* New York, London: John Wiley & Sons, Inc., 1960.
Walker, W. F. *Beginner's Guide to Jig and Tool Design.* New York: Hart Publishing Co., Inc., 1969.

ARTICLES

Dickelman, Ed D. "Flat-Ground Stock Can Lower Tool and Die Costs," *The Tool and Manufacturing Engineer,* June, 1965.
Dittmer, Walter. "Erector Set Cuts Tooling Cost," *American Machinist,* July 5, 1965.
Duckworth, Len. "Magnesium Tooling . . . All Signs Say Go," *The Tool and Manufacturing Engineer,* October, 1968.
Hockett, Jack M. "Putting Magnesium to Work," *The Tool and Manufacturing Engineer,* July, 1961.
Lawry, William A. "Four Ways of Building Epoxy Jigs," *Modern Plastics,* August, 1959.
Mudhar, J. S. "Economic Justification of Jigs and Fixtures," *The Tool and Manufacturing Engineer,* April, 1968.
———. "Graphs Determine Economics of Jigs and Fixtures," *The Tool and Manufacturing Engineer,* August, 1968.
Novak, Lou. "From Short Order Tooling to Short Order Production," *Tooling and Production,* November, 1954.

Sedlik, Harold. "How to Reduce Tool Design Costs," *Modern Machine Shop,* July, 1967.
"Unitized Fixtures on NC Miller Saves 9,000 Tooling Hours a Year," *Metalworking Magazine,* January, 1964.

PAPERS

Anker-Simmons, Ronald S. "Laminated Compressed Wood (LCW) Tooling," *ASTME Paper No. EM63538* (1963).
Arrufat, Walter J. "Plastic Tooling for Small Parts," *ASTME Paper No. EM57138* (1957).
Bertrand, J. L. Q., and C. A. Dawson. "Principles of Fixturing for Machining of Aerospace Materials," *ASTME Paper No. MR66716* (1966).
Bogart, Lewis F. "Wood and Woodlike Products in Tooling," *ASTME Paper No. EM63594* (1963).
―――. "Successful Applications of Plastics to Fixtures," *ASTME Paper No. EM63510* (1963).
―――. "Plastic Fixtures," *ASTME Paper No. CM62506* (1962).
Doule, Lawrence E. "Design and Application of Jigs and Fixtures," *ASTME Paper No. AD62527* (1962).
Dutt, Donald E. "Economics of Tool Selection in Short-Run Production," *ASTME Paper No. MM62522* (1962).
Henderson, Howard V. "Plastics for Tooling," *ASTME Paper No. EM67664* (1967).
Karash, Joseph I. "Drill Jig Design for Secondary Operations," *ASTME Paper No. MR52106* (1952).
London, Arnold. "Cost Savings for Certain Tools Where Plastics are Used Instead of Metal," *ASTME Paper No. MM63600* (1963).
Lyijynen, Fred. "Advantages of Plastic Tooling," *ASTME Paper No. EM56C151A* (1956).
Melde, Karl F. "Magnesium in Aircraft Tooling," *ASTME Paper No. EM58104* (1958).
Munson, Stanley. "Epoxy Plastics – Man-Made Materials for Industry," *ASTME Paper No. EM67131* (1967).
Rupple, Jacob, Jr. "Clamping with Fiberglass Reinforced Plastics," *ASTME Paper No. AD61136* (1961).
Schron, Jack H. "Design and Construction of Economy Jigs and Fixtures," *ASTME Paper No. MS63505* (1963).
Schulte, Sylvester A. "Plastic-Faced-Plaster Spotting Fixtures," *ASTME Paper No. EM67132* (1967).
Sokol, Benjamin. "Plastic Fixtures Have Wide Use," *ASTME Paper No. EM54144* (1954).
Stephany, Harold, and Raymond R. Drake, Jr. "Unit Tooling and Reusable Tooling Systems," *ASTME Paper No. MS62553* (1962).
Taglauer, A. A. "Magnesium – A Light, Versatile Metal for Tooling," *ASTME Paper No. EM67226* (1967).
Tierney, Joseph W. "Plastic Tooling for the Job Shop," *ASTME Paper No. EM62615* (1962).
Young, Allan. "Steels for Jig and Fixture Design," *ASTME Paper No. MS63511* (1963).

MANUFACTURERS' MATERIALS

Brooks and Perkins, Inc. Catalogs and publications.
Devcon Corporation. *Devcon Products for Tooling.*
The Dow Chemical Company. Catalogs and publications.
Kearney & Trecker Corporation. *Fixture Design No. Fix-66.*
Ren Plastics, Inc. *Techniques of Using Epoxy Plastics Tooling Materials.*
Rezolin, Inc. *Manual of Plastic Tooling Materials.*
Wharton Unitools. Catalogs and publications.

index

A

Acetone, 120
Accuracy
  product, 29
  tool, 33, 56, 71, 81–82, 117
Activators, *see* Catalysts
Aerospace industry, 80, 99
Air voids, in plastic, 114, 121, 122, 124
Alcohol, 120
Allen socket nuts, 106, 110
Aluminum, 97, 114, 127
  compared to plastic, 116
  compared to magnesium, 75, 76, 77,
    82–84
Angle iron, 71
Asbestos, 127

B

"Backup" tooling, 64, 65
Birch, 99
Brass, 114
Bronze, 114
Bushing plates, 34, 38–39, 52, 53–54, 56,
    67–68, 108, 110
Bushings, drill, 15, 31, 105, 127–29
Butt joint, 100

C

Capital, investment of, 16, 18–19, 48
Casting, 70
  of plastics, 116, 120, 122–26, 127–29
Casting core, 122–23
Cast iron, 68, 79
  compared to magnesium, 76, 77
Catalysts, 114, 117, 126
Ceramics, 114
Changeover, 48–49
  break-even point, 49
Changeover time, 48
Checking fixtures, 84, 122
Chucks, 35, 126
Clamps, 28, 31, 33, 34, 51, 56, 73
Commercial universal jigs, 34–35
Compressed wood, 98–99
  machinability of, 99
Construction time, 64–65
Converters, *see* Catalysts
Coolants, 76, 78, 79, 116
Cost analysis, 10–13, 48–49
Cost estimates, 48–49
Cotton cloth, 114, 127
Counterboring (recessing), 26
Cutting fluids, *see* Coolants

D

Dado joint, 100–101
Design
  basic tool, 31–34
  plastic tool, 118–120
  principles of, 22–39
  process of, 41
  savings in, 71
  simplification of, 68, 73
  standard commercial tool, 34–39
  tool, 41–46, 50–56, 59
  wood tool, 99–107
Dies, 99, 122
Direct costs, 9, 22
Dovetail joint, 101
Dowel pins, 26, 33, 106
Drill jigs, 23, 29, 67–68, 71, 106, 122
  advantages of, 4, 6, 7, 14
  commercial universal, 34–35
  definition, 3
  indexing, 36
  plate, 33–34, 108
  "sandwich," 34
  template, 31–33
  tumble, 26, 76, 110
Drill rod, 23
Drills, 32

E

Epoxy paste, 126–27
Epoxy plastics
  advantages of, 113, 115–17
  casting of, 122–26
  corrosion resistance of, 116
  costs of, 116
  design with, 118–20
  dimensional stability of, 117, 127
  formability of, 116
  hardware for, 127–29
  hazards of, 117
  heat resistance of, 117
  lightness of, 116, 117
  limitations of, 113, 115–17
  machining of, 126–27
  modifications to, 117, 118
  properties of, 114, 118
  selection of, 118
  strength of, 117, 125–26
  fabrication with, 120–29
"Erector-set" tooling, 22
  adaptability of, 65
  advantages of, 64–65
  components of, 60–61
  construction of, 60–64

"Erector-set" tooling, (*Continued*)
　design of, 59–60
　machining of, 62
　recording of, 64
　reusability of, 65
Extruded forms, 22, 80, 88–90, 92

F

Fabrication time, 23, 68, 116
Fiber-glass cloth, 114, 121
Fillers, 114–15, 124–25, 126, 127
　properties of, 115
Fire-resistant wood, 92
Fixture keys, 26
　patented, reamed-hole, 26–28, 51, 52
Fixtures, 23, 68 (*see also* Checking fixtures,
　Gaging fixtures, Grinding fixtures)
　advantages of, 4, 6, 7, 14
　definition, 3
Formed-section material, 67–73

G

Gages, 16, 122
Gaging fixtures, 81–82
Gel, laminating, 121
Glass, 114, 127
Glues, 100
Granite, 114
Gravel, 114
Grinding fixtures, 68
Groove pins, *see* Spring pins
Growth rings, 93

H

Handling equipment, 76
Hardening agents, *see* Catalysts
Hardwoods, 92, 99
Heat conduction, 76, 79
Heat distortion, 23, 121
Heat resistance, 92, 98, 117
Heat treatment, 22

I

Indirect costs, 9
Inert gases, 78
Inspection, 16, 46
Interchangeability
　product, 4, 6, 18, 80
　tool part, 57
Inventory, 16
Iron oxide, 114
Irregular parts, 109–10, 116

J

Joinery, 100–101
Jig feet, 26

K

Keys, *see* Fixture keys

L

Labor costs, 4, 18, 41
　and tool costs, 9–10, 11
Lacquer, 107, 120
Lamination
　plastic, 116, 120, 121–22, 129
　wood, 96 (*see also* Metal-clad plywood,
　　Plywood)
Lead, 114
Lead time, 5, 57, 110
　reduction of, 64, 68, 116
Limestone, 114
Limited production
　definition, 4
　jigs and fixtures in, 4, 5, 6
　reasons for, 4–5
Limited-production group, 16
Linseed oil, 106–107
Loading, 21–22, 62
Locators, 28, 29–31, 34, 102
Lubricants, 78

M

Machine setup, 14
Machine tables, 26, 37, 51
Machine vises, 36–39
　versatility of, 38–39
Machining, 71
　need for, 62
　reduction of, 21, 22–28
Machining rates, 76–77
Magnesium, 22
　and alloys, 75–76
　characteristics of, 75–80
　costs of, 82–84
　fire protection for, 79–80
　flammability of, 75, 79–80
　forms for tooling, 80–90
　friction with, 78, 79
　machinability of, 75, 76–77, 83
　magnetic properties of, 79
　physical properties of, 76–79, 81
　sparking of, 78–79
　stiffness of, 77
　strength of, 77
　thermal conductivity of, 79
　vibration damping of, 78

Management, 11, 16, 18–19
Manufacturing costs, 9
Manufacturing process sheet, 51
Maple, 99, 110
Mass casting, 124–26
Mass production, 4, 13, 57, 116
Master part system, 14–16
   limitations of, 16
Master pattern, 116, 120–21, 122–25
Material costs
   product, 9
   tool, 83, 114, 116, 125
Materials, tooling, 22, 60
Mechanical reinforcements, 101
Medium production, 7
Melting point, 79
Metal-clad plywood, 97–98
Mill vises, 37–39
Miter joint, 101
Mortise-and-tenon joint, 101
Multipurpose tooling, 47–57, 72–73
   accuracy of, 56
   advantages of, 56–57
   cost advantages of, 48–49
   design of, 50–56
   limitations of, 47

N

Nickel, 97
   compared to magnesium, 77
Numerical control, 7

O

Oil, 78, 116
Operations, similarities in, 47, 48, 50–52

P

Paint, 78, 107
Parting agent, 120–21
Parts, similarities in, 47, 48, 52–56
Patternmakers, 99
Phenolic plastics, 114
Pine, 99, 110
Plaster, 116, 120
Plastic paste construction, 120
Plastics, 22, 113 (see also Epoxy plastics,
       Phenolic plastics, Polyester plastics)
Plate jigs, 33–34, 108
Plywood, 97, 106, 108
Polyester plastics, 114
Prefinished materials, 31
Preproduction runs, 6–7
Pressure pot, 123–24
Processed wood, 96–99
Process engineers, 59

Process operation drawing, 59
Process sheets, 14
Production quantity, 5–6, 47–48
   and tooling cost, 10–11
Production schedules, 47–48
Production time, 4, 18
Proposed tool evaluation form, 17
Prototype parts, 5, 7, 61, 120
Prototype tooling, 65

Q

Quartz, 114

R

Rabbet joint, 100–101
Reinforcements, 118–20, 122
Resins, plastic, 113–14, 126
Rest buttons, 26, 28, 29
Rigidity, 34, 61, 64, 101, 122
Route sheets, see Process sheets

S

Sample parts, 61
Sand, 114
"Sandwich" jig, 34
Savings
   calculation of, 11–12
   money, 5–6, 7–8, 28–31, 33, 34–35,
       49, 65, 70–71, 73, 80, 82–84, 108–10,
       116, 122
   material, 125
   time, 23, 46, 64–65, 73, 122, 125
Screw heads
   as jig feet, 23–26, 73
   as locators, 29–31
Screws, 23–26
   socket set, 56
   with wood, 105
Shellac, 107
Shrinkage, 92, 93, 94–95, 121
Silicon carbide, 114
Single part production, 7
Skill levels, 10, 11, 16, 31
Smothering compounds, 79–80
Softwoods, 92, 99
Space requirements, 56–57
Spare parts, 5, 16
Spring pins, 26, 106
Stainless steel, 97, 114
Standard commercial premachined
       forms, see Tooling plate
Standard components, 22, 28, 57, 60,
       102–106, 127–29
Steel, 22, 80
   cold-rolled, 23

Steel (*Continued*)
    compared to magnesium, 75, 76,
        77, 82–84
    compared to plastic, 116
    compared to wood, 91, 108, 109
    stainless, 97, 114
Stiffness-to-weight ratios, 77
Strength-to-weight ratios, 76, 77
Stress relieving, 78
Structural forms, 71–73
Surface casting, 122–24
Surface finish, 22–23, 77, 81
Swivel vises, 36–37

T

Template drill jigs, 31–33
Templates, 15, 62
Tensile strengths, 77
Thermal expansion, 91
    coefficients of, 118–20
    of magnesium, 78, 80
Thermoplastics, 113
Thermosetting plastics, 113 (*see also*
        Epoxy plastics)
Threaded inserts, 105–106
Tolerances
    material, 23, 77, 81
    product, 21, 26, 29
    tool, 15, 29–31, 56, 117
Tool assembler, 63–64
Tool designers, 41, 42, 43, 99, 118
Tool drawings, 59–60, 118
    methods of, 42–46
    shortcuts in, 46
    simplification of, 68
    symbols and notes in, 22–23
Tool engineers, 11, 16–17, 18, 22,
        23, 28, 29, 47, 59
Tooling
    advantages of, 18–19
    alternatives to, 13–16, 18
    analysis of, 17–18
    cost analysis of, 11–13
    decision making about, 18–19
    definition of, 3
    evaluation of, 16–19
    justification of, 18–19
    multipurpose, 42–57
    principles of, 21–22
    protection of, 78, 106–107
    requirements of, 3–4, 21
    standard, 21
Tooling costs, 4, 5–8, 21, 22, 28, 33,
        34–35, 47, 65, 73, 82–84, 114, 116
    estimation of, 48–49
    and labor costs, 9–10, 11
    minimum, 13
    and product quality, 10–11

Tooling Costs (*Continued*)
    variations in, 10, 11
Tooling estimate, 11–13
Tooling plate, 23, 67–71
    magnesium, 80–81
Toolmakers, 42, 99
T slots, 26, 37, 61
Tumble jigs, 26, 76, 110

U

Universal "erector-set" tooling systems,
        see "Erector-set" tooling
Universal rotary table, 35–36
Universal commercial tooling, 22, 34–39
Unloading, 21–22, 62

V

Vibration damping, 76, 78, 91
Vises, 34–35

W

Walnut shells, 114
Warpage, 92, 94–96, 122
Wax, 120–21
Wear, 22, 97, 102
Wear plates, 102
Weight, reduction of, 75–76, 108,
        110, 116, 117
Welding, 23, 70, 71, 82, 110, 120
    fluxes for, 78
    magnesium, 77–78
Wood, 22 (*see also* Compressed wood,
        Fire-resistant wood, Metal-clad
        plywood, Plywood, Processed wood)
    design with, 99–107
    economics of, 108–10
    expansion of, 96–97
    flammability of, 92
    formability of, 91
    hardness of, 92
    hardware for, 102–106
    lamination of, 96
    for plastic patterns, 116, 120
    properties of, 91–92
    protection of, 106–107
    reinforcements for, 100, 101
    in steel tools, 109–10
    strength of, 91, 92, 93–94, 96, 97, 99
    thermal expansion of, 91
    vibration resistance of, 91
    warpage of, 92, 94–96
Wood sealer, 107
Worker fatigue, 76, 80

Z

Zinc, 75